Biodiversity and Sustainable Development

BIODIVERSITY AND SUSTAINABLE DEVELOPMENT

By
Dr. M. Lakshmi Narasaiah
M.A., Ph.D.
Professor of Economics,
Coordinator, Department of M.B.A.
Sri Krishnadevaraya University Post-graduate Centre,
Kurnool–518 002
Andhra Pradesh (India)

DISCOVERY PUBLISHING HOUSE
NEW DELHI

First Published - 2005

Reprinted - 2025

ISBN: 978-81-7141-945-6

Biodiversity and Sustainable Development

Published by:

DISCOVERY PUBLISHING HOUSE
4383/4B, Ansari Road, Darya Ganj
New Delhi-110 002 (India)
Phone: +91-11-23279245, 23253475; 43596065
E-mail: discoverybooksindia@gmail.com
discoverypublishinghouse@gmail.com
web: www.discoverypublishinggroup.com

Printed at:
Infinity Imaging Systems
Delhi

PREFACE

As the century begins, natural resources are under increasing pressure, threatening public health and development. Water shortages, soil exhaustion, loss of forests, air and water pollution, and degradation of coastlines afflict many areas. As the world's population grows, improving living standards without destroying the environment is a global challenge.

Most developed economies currently consume resources much faster than they can regenerate. Most developing countries with rapid population growth face the urgent need to improve living standards. As we humans exploit nature to meet present needs, are we destroying resources needed for the future?

Environment Getting Worse

In the past decade in every environmental sector, conditions have either failed to improve, or they are worsening:

Public Health. Unclean water, along with poor sanitation, kills over 12 million people each year, most in developing countries. Air pollution kills nearly 3 million more. Heavy metals and other contaminants also cause widespread health problems.

Food Supply. Will there be enough food to go around? In 64 of 105 developing countries, studied by UN Food and Agricultural organisation, the population has been growing faster than food supplies. Population pressures have degraded some 2 billion hectares of arable land—an area the size of Canada and the US.

Fresh Water. The supply of freshwater is finite, but demand is soaring as population grows and use per capita rises. By 2025, when world population is projected to be 8 billion, 48 countries, containing 3 billion people will face shortages.

Coastlines and Oceans. Half of all coastal ecosystems are pressured by high population densities and urban development. A tide of pollution is rising in the world's seas. Ocean fisheries are being overexploited, and fish catches are down.

Forests. Nearly half of the world's original forest cover has been lost, and each year another 16 million hectares are cut, bulldozed, or burned. Forests provide over US$ 400 billion to the world economy annually and are vital to maintaining healthy ecosystems. Yet, current demand for forest products may exceed the limit of sustainable consumption by 25 per cent.

Biodiversity. The earth's biological diversity is crucial to the continued vitality of agriculture and medicine, and perhaps even to life on earth itself. Yet human activities are pushing many thousands of plant and animal species into extinction. Two of every three species is estimated to be in decline.

Global Climate Change. The earth's surface is warming due to greenhouse gas emissions, largely from burning fossil fuels. If the global temperature rises as projected, sea levels would rise by several meters, causing widespread flooding. Global warming also could cause droughts and disrupt agriculture.

Dr. M. Lakshmi Narasaiah

CONTENTS

1

BIODIVERSITY

As human population has surged this century, the populations of numerous other species have tumbled, many to the point of extinction. Indeed, we live amid the greatest extinction of plant and animal life since the dinosaurs disappeared some 65 million years ago, with species losses at 100 to 1,000 times the natural rate. But humans are not just witnesses to a rare historic event, we are actually its cause. The leading sources of today's species loss, habitat alteration, invasions by exotic species, pollution, and overhunting are all a function of human activities.

Human activities have pushed the percentage of mammals, amphibians, land fish that are in "immediate danger" of extinction into double digits. The principal cause of species extinction is habitat loss—the result of encroachment by humans for settlements, for agriculture, or to claim resources such as timber. A particularly productive but vulnerable habitat is found in coastal areas, home to 60 per cent of the world's population. Coastal wetlands nurture two thirds of all commercially caught fish, for example. And coral reefs have the second highest concentration of biodiversity in the world, after tropical rainforests. But human encroachment and pollution are degrading these areas: roughly half of the world's salt marshes and mangrove swamps have been eliminated or radically altered, and two thirds of the world's coral reefs have been degraded, 10 per cent of them "beyond recognition". As coastal migration continues—coastal dwellers could account for 75 per cent of world population within 30 years—the pressures on these productive habitats will likely increase.

Habitat loss tends to accelerate with an increase in a country's population density. This is bad news for the world's biodiversity hotspots-species-rich ecosystems at greatest risk of destruction. Twenty-four of these hotspots, containing half of the planet's species, have been identified globally. Some of the most important hotspot countries will reach population densities that have been linked with very high rates of habitat loss. Five of the six most biologically rich countries could see more than two thirds of their original habitat destroyed by 2050 if this historical relationship holds.

Related to loss of habitat is the growing incidence of plant, animal, insect, and microbial invasions of ecosystems worldwide as human interchange increases. These "exotic species" sometimes dominate local ecosystems, eliminating native species and reducing overall diversity. Exotics are implicated in 68 per cent of all fish extinctions in the United States this century, for example. Growth in human travel and commerce explains many accidental invasions by exotics, but foreign species are also deliberately introduced into farms, plantation forests, and aquaculture systems. Although only 1 per cent of exotics cause widespread damage, exotic species are the second leading cause, after habitat destruction, of species loss worldwide.

Other, often diffuse effects of expanded human activities also disrupt ecosystems. Nitrogen, for example, is now made available to plants at more than twice the preindustrial rate as a result of fertilizer production, cultivation of nitrogen-fixing crops, and the burning of fossil fuels. This overfertilisation of the Earth favours some species at the expense of others, leading to a reduction in diversity and resiliency of land and aquatic ecosystems.

Likewise, greenhouse gas emissions could disrupt ecosystems on a vast scale. As with nitrogen, increased levels of atmospheric carbon may favour some species over others:

annuals over perennials, for example, or deciduous trees over evergreens. To the extent that greenhouse gases induce changes in global climate, many species may be at risk as habitats shift or shrink, and as some life forms, such as insects or animals, adapt and migrate more quickly than others, such as plants. And as sea levels rise with a change in climate, ecosystems such as coastal wetlands could be destroyed.

2

LIVING WITH DIVERSITY

Fishers' nets and loggers' saws may directly impoverish local ecosystems, but most biological losses have root causes far away, in long-settled urban areas and farms where diversity is seldom a concern, but where steadily rising demand for food, water, wood and other resource—and the dispersal of resulting wastes—reach far beyond the settled areas themselves. In general, these peopled landscapes have lost much of their own biological wealth, but what remains is still important to their continued functioning and livability. Reconciling farms and cities with diversity will require stopping the damage they bring to remaining natural habitats, but also beginning to halt and reverse the homogenisation of these unnatural habitats.

Uniformity is not inherently undesirable. In fact, to some degree, homogeneity is the basis of all agriculture: a given type of plant is favorued and others are suppressed or eliminated. But trends in recent decades (most notably the Green Revolution and the parallel intensification of farming systems in industrial nations) have pushed uniformity to dangerous levels.

The unsustainability of modern agriculture is in part a measure of its inability to tolerate diversity. Both genetic and ecological uniformity—the sameness of fields sown horizon to horizon without interruption—demand costly and often futile reliance on chemicals to protect crops from pests or diseases that are rapidly spreading and evolving. The drive to leave no hectare unplowed worsens soil erosion, pushing tractors onto highly erodible hillsides and removing windbreaks, hedgerows and other remnant habitats.

Some of agricultur's biological impacts are obvious—the expansion of farms onto forests and wetlands, for example. While the increasing reliance on chemical inputs and machinery has reduced these impacts in some cases by decreasing the area needed to produce a given amount of food, it has worsened others.

A fundamental transition away from today's wasteful and polluting farming systems is needed to put the world's food supplies on a secure footing. Many of the reforms that will reduce farming's dependence on fossil fuel inputs and its misuse of soils and waters can also restore diversity to agricultural landscapes. Pesticides, for example, kill not only pests but other animals, such as pollinators and predators, that are beneficial to agriculture. Alternative pest control measures that lower pesticide use can also, ironically, reduce pest damage to crops by reviving the diversity of soil and insect communities, which play crucial roles in maintaining soil productivity and checking the spread of pest outbreaks.

Traditional agroecosystems are important not only because they provide sustenance to rural people and harbour valuable genetic resources, but also because they contain the seeds of a sustainable, diversity-based mode of agriculture. At varying levels, diversity is the basis of production for many peasants. Farmers often mix strains of a given crop in their fields as a hedge against the vagaries of weather. They also tend to recognize the dependence of their farms on adjacent ecological systems and to tolerate wild plants (often crop relatives whose continued interbreeding with domestic descendants contributes to genetic variety) on the outskirts of their fields.

Population growth and the expansion of large commercial farms have rendered many once-sound practices no loner viable, and traditional agriculture badly needs infusions of money and research to increase its modest yields without abandoning its stability.

Urban areas, with good reason, are considered the antithesis of natural diversity. Only the most resilient creatures (many of them regarded as weeds and pests) thrive in them, and cities' ceaseless expansion, consumption of resources, and emissions of waste threaten both farmland and wilderness almost everywhere. As with agricultural lands, the first priority for urban areas is to half their expansion onto other ecosystems and reduce the damage they export, such as the sewage poured onto coral reefs by burgeoning coastal cities throughout the tropics, or the wasteful consumption of tropical hardwoods in Japanese building construction.

But even concrete jungles can support some diversity. Landscaping of private yards and public spaces with native vegetation can not only reduce the expense and environmental impact of watering, spraying and hauling the remains of sterile grass monocultures, but also help revive bird and other wildlife populations. Most urban areas also have water-ways running through them, or corridors of unused land such as steep ravines; if their use as waste receptacles is reduced, these can be maintained or restored as wildlife habitat.

In developing nations, especially, a surprising amount of agricultural production takes place within city limits, in home gardens. These hidden farmlands contain a great deal of genetic diversity, and their expansion could help reduce the scale and environmental impacts of commercial agriculture.

One reason that the destruction of biological diversity has gone so far without major public commitments to stopping it is that urban dwellers have little experience of the natural and even less understanding of its importance. Restoring nature where people live-reestablishing a personal link with the living world—may be necessary to save it elsewhere. For all the rational arguments favouring long-term protection of biological assets, people who have lost all direct sense of their dependence on natural systems may simply not care.

Only a growing respect for diversity for its own sake—beginning, perhaps, with a reconnection between people and nature within the urban environment—will trigger altruistic responses among those wealthy enough to have the option of considering the needs of future generations and natural communities. Although many conservation measures make economic sense, arguments of economics or self-interest will likely fail to be convincing when the contest is between a few uncharismatic species of unknown value and a major industrial project. "Human beings make sacrifices for what they love." Those who maintain strong bonds with the biological world on which they depend may be more inclined to make the hard decisions needed to protect it.

3

SAVING THE PLANET:
Imperialism in Green Garb?

Developing countries feel that protecting the world's resources is just another way for rich nations to retain the upper hand in the international trade game. For nearly a decade, international efforts to address global environmental concerns have been frustrated by a deep rift in perceptions between rich and poor countries. Economists and environmentalists in developing nations argue that the North almost exclusively drives the agenda for environmental negotiations. Under the pretext of saving the planet, they say, the industrialized world is wielding a new brand of dominance, "ecoimperialism".

Developing countries like India and China continue to resist global environmental protocols, like the 1989 Montreal Accord to cut the production of CFC gases (used for example in refrigerators) by 50 per cent, or the Clean Development Mechanism (CDM), part of the climate change negotiations initiated under the 1997 Kyoto Protocol.

The spectre of imperialism is likely to vitiate the next round of climate change talks in Bonn (Germany) when policymakers finalize the terms on which the CDM will be implemented. Negotiated by industrialized countries to gain some flexibility in meeting the emission reduction targets pledged in Kyoto, few issues in recent environmental diplomacy are proving as contentious.

Critics say the mechanism is the latest in a string of attempts to dominate poor countries, which are being virtually "bribed" so that rich nations can continue business as usual.

By financing forestry schemes and other energy-efficient projects, industrialized countries could exploit the mechanism to avoid reducing their own greenhouse gases. Environmentalists fear this could turn the Amazon and other primeval forests into "carbon sinks" to absorb pollution, but with side effects which disregard the development needs of southern countries.

Lopsided Negotiations

The Northern bias continues to dominate discussion of the global climatic crisis. The threat to the atmospheric commons has been building over centuries mainly because of industrial activity in the North. Yet discussions seems to focus more on developing countries; the North refuses to assume extra responsibility for cleaning up the atmosphere. No wonder the Third World cries foul when it is asked to share the costs.

The 1989 Basel Convention, for instance, imposed restrictions on trade in scrap metals and recyclable materials, claiming they were hazardous to the environment. Economists say it prohibits poor countries from competing in the lucrative world market for computer parts, scrap metal, and recyclable products.

Other examples of trade restrictions are cited. In the early 1990s, Malaysia and Indonesia fought to overturn and ecolabelling law introduced by Austria ostensibly to safeguard the Asian rain forests. Austria refused to import timber that was not from sustainable managed forests, but no such curbs existed for wood from temperate areas. The protectionist flavour of the measure was overt, and Austria eventually revoked it.

In other trade-environment disputes over the last decade, the United States has been accused of protectionism in banning the import of Mexican tuna because dolphins were getting ensnared and killed in nets meant for the fish. Shrimp from India, Pakistan, Thailand and Malaysia, which paid no heed to seaturtle protection, were similarly banned in 1996. The sanctions may have been motivated by a desire to protect dolphins and turtles, but the poorer countries claimed that they were a pretext for suppressing competition in the global fish market.

A Green Agenda to Stop Growth?

Rules restricting trade through the Basel Convention or attempts to ban genetically modified foods are designed to exclude poorer countries from world markets. Other voices in the South, however argue that environmental controls like the CDM are not all bad. Developing countries, say experts, will receive $5 to $17 billion to fund climate-friendly technologies. The CDM gives us an opportunity to invest in projects that promote sustainable development. If incidentally they also reduce emissions, we should't quarrel with the fact.

4

POPULATION AND THE ENVIRONMENT: *The Global Challenge*

As the century begin, natural resources are under increasing pressure, threatening public health and development. Water shortages, soil exhaustion, loss of forests, air and water pollution, and degradation of coastlines afflict many areas. As the world's population grows, improving living standards without destroying the environment is a global challenge.

Most developed economies currently consume resources much faster than they can regenerate. Most developing countries with rapid population growth face the urgent need to improve living standards. As we humans exploit nature to meet present needs, are we destroying resources needed for the future?

Environment Getting Worse

In the past decade in every environmental sector, conditions have either failed to improve, or they are worsening:

Public Health. Unclean water, along with poor sanitation, kills over 12 million people each year, most in developing countries. Air pollution kills nearly 3 million more. Heavy metals and other contaminants also cause widespread health problems.

Food Supply. Will there be enough food to go around? In 64 of 105 developing countries, studied by UN Food and Agricultural organisation, the population has been growing faster than food supplies. Population pressures have degraded some 2 billion hectares of arable land—an area the size of Canada and the US.

Fresh Water. The supply of fresh water is finite, but demand is soaring as population grows and use per capita rises. By 2025, when world population is projected to be 8 billion, 48 countries, containing 3 billion people will face shortages.

Coastlines and Oceans. Half of all coastal ecosystems are pressured by high population densities and urban development. A tide of pollution is rising in the world's seas. Ocean fisheries are being over-exploited, and fish catches are down.

Forests. Nearly half of the world's original forest-cover has been lost, and each year another 16 million hectares are cut, bulldozed, or burned. Forests provide over US$400 billion to the world economy annually and are vital to maintaining healthy ecosystems. Yet, current demand for forest products may exceed the limit of sustainable consumption by 25 per cent.

Biodiversity. The earth's biological diversity is crucial to the continued vitality of agriculture and medicine, and perhaps even to life on earths itself. Yet human activities are pushing many thousands of plant and animal species into extinction. Two of every three species is estimated to be in decline.

Global Climate Change. The earth's surface is warming due to greenhouse gas emissions, largely from burning fossil fuels. If the global temperature rises as projected, sea levels would rise by several meters, causing widespread flooding. Global warming also could cause droughts and disrupt agriculture.

Toward a Livable Future

How people preserve or abuse the environment could largely determine whether living standards improve or deteriorate. Growing human numbers, urban expansion, and resource exploitation do not bode well for the future. Without practicing sustainable development, humanity faces a deteriorating environment and may even invite ecological disaster.

Taking Action. Many steps toward sustainability can be taken today. The include using energy more efficiently; managing cities better; phasing out subsidies that encourage waste; managing water resources and protecting fresh-water sources; harvesting forest products rather than destroying forests; preserving arable land and increasing food production through a second Green Revolution; managing coastal zones and ocean fisheries; protecting biodiversity hotspots; and adopting an international convention on climate change.

Stabilizing Population. While population growth has slowed, the absolute number of people continues to increase by about 1 billion every 13 years, slowing population growth would help improve living standards and would buy time to protect natural resources. In the long run, to sustain higher living standards world population size must stabilize.

5

ECOSYSTEMS, OUR UNKNOWN PROTECTORS

How ecosystems work and what part they play in biodiversity remain a mystery. But we do know that they perform a host of invaluable services for the human species. In my view, biodiversity's fundamental value is neither aesthetic nor economic but environmental, even though most people are largely unaware of this. The value of biodiversity is often measured in terms of the number of species living in a given area. But the interactions between the many species in an ecosystem, and between them and the environment's physical and chemical components are also very important. This highly intricate web of relationships makes an ecosystem more valuable than the sum of the species it contains.

Ecosystems perform services that are essential for the survival of the human species. They fix carbon in the atmosphere and produce oxygen, protect soil from erosion and keep it fertile, filter water and replenish aquifers, provide pollination and anti-parasite agents and so on.

The first two of these services are closely related to each other. They result from photosynthesis, whereby green plants, starting with algae, absorb carbon dioxide (CO_2) and emit oxygen. For millions for years, the balance between the various gases in the atmosphere remained stable. But with the coming of the industrial revolution, humans began burning increase amounts of fossil fuels. Today, three billion tonnes of carbon build up in the atmosphere each year and natural ecosystems can no longer absorb all these emissions—especially since they are disappearing at an alarming rate.

Deforestation alone releases such tremendous amounts of CO_2 and other gases, such as methane, that it has become the second-leading cause of global warning.

Strong fresh-water, protecting soil and keeping it fertile are three other closely related functions. Ecosystems are veritable "freshwater factories". They absorb rainwater and slowly filter it through the soil before draining it towards streams, rivers, lakes and underground aquifers that supply us with the precious liquid. When the vegetal ground cover is degraded, the water cycle is disrupted. Rain strikes the bare earth, washing away huge amounts of nutritional substances. Reservoirs, lakes and rivers silt up.

Uncertain Reaction to Climate Change

Despite years of research, scientists still know very little about how ecosystems work. We are generally incapable of predicting how they will react to certain transformations in the environment, especially climate changes. Nor do we know any more about whether a species present in a given environment is superfluous or "replicable", even when it is very rare. Likewise, we do not know which key species are indispensable to maintaining an ecosystem, with a few exceptions such as pine forests, where that tree is obviously the dominant species.

We know even less about the part biological diversity itself play in maintaining ecosystems and the services they perform. One simple example is a highest diversified forest that absorbs carbon dioxide—a vital function, as we have seen, for limiting global warming. Suppose the forest is cleared to make way for a single-crop forest. The service will still be performed, perhaps even better at first because young, fast-growing trees absorb more CO_2 than old ones, which regenerate slowly. But what will happen in the long term? After several decades, the consequences of the loss of biodiversity will probably be felt. Replacing many species with a single one will have certainly depleted the soil and, in the long term, slowed down the forest's growth and consequently its ability to absorb CO_2.

More generally, diversified ecosystems seem more productive. Specialists remain wary about their conclusions, but today they believe that biodiversity helps ecosystems to resist alien species and diseases and to recover faster in the event of disruption. In the face of doubt, and to find out more about them, it is better to preserve as many different ecosystems as possible.

A Costly Lesson for New York City

Most people take it for granted that ecosystems will carry on performing services without receiving anything in return. They think nature will continue benefiting humanity, no matter how much damage is done. The survival of organisms other than our own species is perceived as a frill that future generations can live without.

These preconceived ideas are wrong and dangerous as the city of New York has recently come to realize. The city's water has always enjoyed such a good reputation that it was sold throughout the northeastern United States. Its equality was due to Catskill Mountains' natural purification system. But that ecosystem has suffered so much from pollution, especially fertilizer run-off from farms, that by the late 1990s New York's water had become undrinkable. The city planned to build a purification plant, whose cost was put at between six and eight billion dollars, not including the $300 million in yearly operating costs—an astronomical bill for a service that had always been free! The price was so staggering that the city eventually decided to restore the Catskill Mountains' degraded environment at a cost of only one billion dollars.

This story clearly illustrates where our interests lie. We must preserve ecosystems and the conditions that enable our planet to ensure the survival of *Homo sapiens* or, at least, the short-term maintenance of our current quality of life.

6

CLIMATE CHANGE AND HUMAN HEALTH

Changes in the India's climate, stemming from the greenhouse effect, are highly likely to damage human health. Food and fresh water supplies will be disrupted, millions of people displaced, and disease patterns altered dangerously and unpredictably.

Human health could be affected by even quite small changes in average mean temperature, and there is the prospect of some major diseases flourishing in warmer conditions and of more resistant strains of infection emerging.

The population in India most vulnerable to the negative impacts of global warming are in the lower-income groups, residents of coastal lowlands and islands, those living in semi-arid lands, and the urban poor in the squatter settlements, slums and shanty-towns of large cities.

Present strategies for immunisation, coping with disease vectors or carriers, providing safe drinking water, and improving nutrition are all based on existing climate regimes, ecosystems, and sea and solar radiation levels. These are all expected to change, but exactly how much cannot be predicted. It is therefore, virtually impossible to adjust health and nutritional strategies to take account of possible climate changes.

Humans can adapt to moderate changes in temperature and to occasional extremes. But this adaptive capacity is relatively low in infants and the elderly; it rises through childhood and adolescence to reach a maximum which can be maintained up to about 30 years of age.

A changing climate would alter the ecosystems of the vectors or agents which carry or cause many diseases, whether these be viruses, bacteria, parasites, plants, insects or other animals such as mosquitoes and snails. As the weather warms, the boundaries of the tropics may extend into the present subtropics, and parts of temperate areas may become subtropical. As air temperatures increase, some diseases will become common in regions which once rarely know them and where there is little natural resistance to them. As a result, death rates may also climb significantly.

It is possible that warmer weather around the world will cause increases in summer diseases and decreases in those associated with winter. Diseases contracted from both water and air will also spread more readily as ambient temperatures rise. In a warmer climate, mosquitoes and other vectors also may migrate vertically, up into highlands which were once too cold for them. This may be particularly hazardous in tropical highland areas where there is no natural resistance to malaiara.

Changes in temperature, rainfall, humidity and storm patterns may affect diseases borne by vectors in two ways. First, they will directly affect the vector's reproduction rate, biting rate, and the duration and frequency of human exposure. Second, they may modify agricultural systems or plant species; thus changing the relationship between host and vector. Development rates of malarial mosquitoes, for example, increase with warmer temperatures, but these pests need wet areas in which to breed.

Sea-level rise could also spread infectious disease by flooding sewerage and sanitation systems in coastal cities, and increase the incidence of diarrhoea in children. The flooding of hazardous waste dumps and sanitation systems could lead to long-term contamination of crop lands.

Rising seas may also disrupt marine habitats land aquatic food chains. Since fish constitute 40 per cent of all animal protein consumed by the people of India such a disruption of the marine ecosystem would affect the food supplies of many millions of people and dramatically increase protein deficiency and malnutrition. Changes in the availability of food and water, as well as radical shifts in disease patterns, could initiate large migrations of people, exacerbating food shortages, overcrowding, social stress and instability.

Some of the factors contributing significantly to global warming, such as the burning of fossil fuels and the use of chlorofluorocarbons (CFCs) and halons, threaten human health in other ways too. A typical petrol-driven motor car, for example, emits carbon monoxide, sulphur and nitrogen oxides, hydrocarbons, low-level ozone and lead-all of which are hazardous to health.

The ozone-depleting CFCs and halons pose a particular threat to humans through an increased risk of skin cancer, cataracts and lower immunity to other illnesses as a result of increased exposure to ultraviolet B radiation from the sun. Skin cancer risks are expected to rise most among fair-skinned.

7

POPULATION GROWTH AND CLIMATE CHANGE

Over the last half-century, carbon emissions from fossil fuel burning expanded at nearly twice the rate of population, boosting atmospheric concentrations of carbon dioxide, the principal greenhouse gas, by 30 per cent over preindustrial levels. All major scientific bodies acknowledge the likelihood that climate change due to the buildup of greenhouse gases in the atmosphere is indeed under way. The 15 warmest years on record have all occurred since 1979, and 1998.

The destabilisation of our climate threatens more intense heat waves, more severe droughts and floods, more destructive storms, and more extensive forest fires. The related shifts in rainfall and temperature may jeopardize food production, the Earth's biological diversity, and entire ecosystems, as well as human health by expanding the ranges of tropical diseases. Unless efforts to curb them are stepped up, carbon emissions will continue to grow faster than population over the next 50 years, driving the earth's climate system into unchartered territory. The Intergovernmental Panel on Climate Change (IPCC) estimates that an eventual two-thirds reduction in global emissions is needed to avoid precariously high levels of atmosphere carbon dioxide concentrations.

The IPCC and U.S. Department of Energy (DOE) project that emissions from developing countries will nearly quadruple over the next half-century, while those from industrial nations will increase by 30 per cent. Although annual emissions from industrial countries are currently twice as high as from developing ones, the latter are on target to eclipse the industrial world by 2020.

Higher per capita carbon emissions accounts for roughly 55 per cent of the increase in emissions projected for developing nations. Emissions per person are due to more than double from 0.51 tons of carbon per year in 2000, just one fifth of the industrial level, to 1.14 tons in 2050. The remaining 45 per cent of emissions increases is due to population growth.

Fossil fuel use accounts for roughly three quarters of world carbon emissions. As a result, regional growth in carbon emissions tend to occur where economic activity, and related energy use, is projected to grow most rapidly. Emissions in China are projected to grow over three times faster than population in the next half-century, as emissions per person soar from 0.77 tons of carbon to 2.81 tons due to booming economy that is heavily reliant on coal and other carbon-rich energy sources. In Africa, in contrast, emissions per person are expected to scarcely change—growing from the current level of 0.33 tons in 2050, despite a threefold increase in total emissions.

The effects of population growth are most profound in countries where people are heavily emitters. For example, the 115 million people added to the population of the United States between 1950 and 1998—an increase of nearly 75 per cent in just 45 years—account for more than one tenth of current global emissions. And the carbon emissions of the 75 million people who will be added to the U.S. population in the next 50 years roughly equal the emission of the 1.3 billion people who will be added to Africa during that period.

Deforestation and other landuse changes account for the remainder of world carbon emissions. Forests have served as a sink for carbon throughout much of human history. In recent years, however, the world's forests have become net sources of atmospheric carbon, largely due to forest burning and clearing in the tropics. Six months of fires in Asia in 1997 and 1998 released more carbon than Western Europe emits from fossil fuel burning in an entire year. The carbon contribution from this source will likely increase in coming years as the burgeoning human population continues to cut down forests.

8

ECONOMICS AND SUSTAINABLE DEVELOPMENT

Economists and ecologists were once seen as enemies: environmental protection, it was thought, could only be achieved at the expense of economic growth. The misconception persists at the extremes among both the most fundamentalist Greens and the most ideological free marketers. But increasingly it is now being recognised that development and care for the environment go hand in hand. This interdependence is coalescing in the new and necessary discipline of environmental economics.

Conventional economics patterns have often assumed that growth and technical progress will nullify all resource and environmental limits. Environmental economics recognises that the world's natural capital underpins all development, and that it is rapidly becoming scarcer as human demands exceed the globe's long-term carrying capacity. Government of India has introduced environmental measures over the last two decades, but need to move further towards integrating them into economic policies. There can be no real sustainable development unless environment and development policies are integrated at the very beginning of the decision-making process.

Quantifying the Environmental Cost

One of the first steps is to work out the true costs of polluting and depleting the World's natural resources, such as its soil, air and water, the climate and the ozone layer. There have often been regarded as free goods, and it was believed that the world has an infinite capacity to absorb the effects of human activities. Environmental economists, recognising that the social and economic costs of degradation are very great, are trying to quantify them. They say that this will make possiblė better use of such tools as cost-benefit analysis, environmental impact assessment and risk assessment—and the production of national income accounts which reflect the depletion and degradation of natural resources. As these costs are identified and quantified, economic policy can increasingly be developed with sustainable development as the primary objective. Achieving sustainable development requires industrialized and developing countries to make dramatic changes in national and international policies based on a global partnership. The greenhouse effect, the destruction of the ozone layer, the extinction of species and contamination of the oceans, and other environmental problems, affect us all, no matter which corner of the globe we inhabit.

The first and essential step in overcoming a difficulty is to recognize it and understand it. Concern over the difficulties related to sustainability has led scientists and national and international institutions to study the concept and suggest ways of meeting its many requirements. Indicators have been established to measure pollution levels, soil erosion, salinisation, deforestation and a host of environmental problems. Evaluating the impact of such natural resource-use on ecosystems is a major step towards finding the necessary solutions.

For example, it has become clear, on a macroeconomic level, that national accounting systems fail to reflect these effects adequately. Deterioration of the world's rivers, land degradation, air pollution and contamination of the seas are not taken into consideration. Inadequate accounting distorts reality and gives a false idea of the consequences of growth and production.

On a microeconomic level, much is being done to redefine production costs. Incorporating the cost of waste management an internalizing negative external impacts within production prices are beneficial aspects of the economics of sustainability.

Steps are being taken to evaluate public and commonly held assets and to put a price on them, even though they may not be subject to market forces. These are only in the earliest stage but they will allow for more accurate evaluation of the world's natural capital. Fiscal, market, quota and other instruments are being developed to enforce change in the way in which certain resources are used. Examples include markets for transferable emission quotas or compensatory taxation mechanisms designed to ensure that economic forces act to reduce greenhouse gas emission. Efforts at impact analysis—and in a general sense, cost-benefit analysis—permit rough estimations of the impact that projects might have no ecosystems.

Long-term Repercussions

These instruments carry significant limitations but they are important nevertheless because they attempt to quantify impacts on the natural world and to achieve a more rational use of natural resources. The development of such instruments and evaluation techniques will have significant repercussions in the formulation of sustainable long-term policies. But we must bear in mind that sustainability is not just an economic issue: it is also a political and cultural one.

The concept of sustainability demands as alternative view point in which humankind and the natural world are perceived as a unit—as different yet mutually sustaining aspects of a whole. This perception is not incompatible with progress. It does not renounce development. It simply seeks to affirm life and refuses to discriminate between the means and the end.

It understands that happiness cannot be achieved by destructive means. The questions of how to produce and how to consume therefore become extremely important. Neither should be at the expense of the future or of the natural world. Efficiency is not limited to the links between investment, products and prices: it must address the rational use of resources, including environmental and cultural consequences, both in the long and the short-term.

Very considerable adjustments must be made in the interests of sustainable development. They demand a reassessment of all our activities which cannot, logically, be done overnight. It is a long continuous process, characterized by steadfastness and compromise.

9

URBANISATION AND THE ENVIRONMENT

Is abandoning the cities the answer to the growing ecological problems of urbanisation? The trend at any rate is in the opposite direction. At the beginning of this century, only every 10th person worldwide was a city dweller. At its end, more than half the global population will be urbanites. And most of the urban population growth will take place in the developing countries, led by Asia.

Compared to other parts of the world, however, the urbanisation process in Asia is currently not even particularly far out in front. Worldwide, city dwellers account for 43 per cent of the total population. Industrial nations have an average urbanisation rate of 72 per cent. Less industrialized countries have 34 per cent. In the Asia-Pacific region the rate is 30 per cent, in Latin America 72 per cent, and in Africa 33 per cent. The urbanisation growth rate in a number of Asian countries has, in fact, slowed compared with earlier years. Nevertheless, not only industrialisation, but also the increasing degree of urbanisation has emerged as a growing burden on the environment in many Asian countries.

Changed Urbanisation Pattern in India

Environment burdens are just as much a problem in the old industrial nations as they are in India. But each group has a specific pattern of development. The urbanisation process in India has proven to be more pollution-intensive than that in the old industrial nations of Europe and North America. There are several reasons for that:

- industrialisation in India is restricted to a few locations which are often concentrated in and around capital cities. Although environmental damage continues to be minor at a national level, these locations have higher pollution levels than those ever reached in Industrial nations;
- furthermore, besides the strong regionalisation of industries, the industrialisation pattern of India shows a great diversity of environmental hazards. The trend to establish "last industries first", which is promoted by progressive industrialisation, leads to a country producing certain dangerous materials before they have been covered by state regulations;
- the time factor has to be seen as an important element in the emergence of these already highly regionalized environmental burdens. In India industrialisation and its concomitant urbanisation is taking place within a ban population grew tremendously.

Growing Environmental Damage

Water pollution in India is caused mainly by domestic sewage. For example, households are responsible for 75 per cent of the pollution of the rivers. The domestic sewage problem got more and more out of control with growing urban population. Pipe-based waste water systems are rare in this country. In India dealing with waste has an extremely low priority. The type of waste disposal depends mostly on what the cities can afford. The present level of air pollution is also very high.

Innovative Approaches to Solutions

Environmental protection and economic development are seen as contradictions. Economic development can only be achieved at the cost of higher levels of environmental pollution. And in reverse, if pollution is to be controlled and reduced this can only be done to the disadvantage of further development. In the meantime, however, numerous instances of successful urban environmental management are developing. They could help to change and subsequently break through the existing pattern of thinking. The following approaches can be viewed as important.

Combining regulations with incentives: The introduction of lead-free petrol and the mandatory equipping of new cars with catalytic converters is still by no means common in India. As numerous cars without catalytic converters are still able to use lead-free. Converters were then at first made compulsory for higher-powered cars, and later also for compact models.

Combining regulations with simple controls: Apart from general limitation of the number of cars in the city, its most important single measure to prevent traffic jams and the additional petrol consumption and pollutant emissions caused by them.

High economic growth in India has, in fact, led to a general reduction of poverty. But the distribution of income, particularly between urban and rural areas, has remained relatively constant. Urban environmental and traffic problems have increased heavily during the same period. These developments can be attributed to a certain pattern of official action (or "non-action"):

- governments have made efforts in supplying roads, but neglected the demand for mobility.
- governments are preoccupied with supplying water, and have neglected follow-up problems, above all the questions of waste water disposal and treatment. In Indian cities, for example, this leads to the absurd situation that due to the mushroom-like growth of the cities and the increased water pollution linked with it, water must be brought in over ever greater distances and at ever greater expense;
- governments take a one-sided look at noxious substances. Concentrations of harmful substances in water and in the air are, in fact, checked, and some measures are taken against individual pollutants of single sectors (e.g., lead emission by the transport sector).

But an integrated policy which operates integrated environmental management with the aim of comprehensively relieving the burdens on the environment has not yet been developed anywhere. To consider such a concept, it is necessary to cut loose from the customary way of approaching problems. It makes sense not to separate the problem areas from each other according to sectors and pollutants, but rather on the basis of their ecological impact.

Orienting on demand hits the core of the concept of ecological modernisation, which is about reducing the intensity of resource-use (note, at this stage this does not yet mean the absolute reduction of inputs). At the same time, sights are set on a lower use of land with the same size of population, or also lower energy consumption with the same degree of added value or the same per capita income.

Finally, the importance of governments for creating framework conditions must be emphasised once again. Because the actors come from different spheres, such conditions are essential.

From the time of the Greek polis, it was the ambition of the Greek city councillors to pass on a city that was more beautiful than the one they had taken over. There is a long way to go before such an attribute asserts itself in India (and elsewhere).

10

SUSTAINABLE CITIES

Today almost one-half of the world's population lives in cities. The world's cities are growing by one million people each week. Cities today play a significant role in development. They continue to attract migrants from rural areas because they enable people to advance socially and economically. Cities offer significant economies of scale in the provision of jobs, housing and services, and are important centres of productivity and social development.

However, the stress of this rapid urban population growth is often overwhelming. The long list of afflictions includes urban poverty rates of up to 60 per cent. Despite growing investments, more than one-third of the urban population live in substandard housing. Forty per cent of urban dwellers do not have access to safe drinking water or adequate sanitation. Primarily due to a rapid growth and a deteriorating urban environment, at least 600 million people in human settlements (cities, towns and villages) already live in health-and life-threatening situations, and almost 50 per cent of these are children.

The high rate of urban population growth in most regions has led to common problems: congestion, lack of funds to provide basic services, a shortage of adequate housing and declining infrastructure, to name a few.

While these problems are occurring in urban areas, cities still have an important role to play in protecting the global environment in the face of rapid urban population growth.

Agricultural and livestock production in rural areas are pushing farther and farther into ecologically fragile regions and cannot support growing population. The finite land and water resources make it imperative that human settlements be carefully planned. Indeed, sustainable urbanisation will ease the pressures caused by encroachment on fragile natural habitats.

India's cities offer a bewildering sight to any visitor: the congestion caused by rapid population growth and a continuing rural-urban drift often leads to conditions which defy all rules of orders, hygiene and environmental safety. Inadequate leadership, corruption and mismanagement have a harmful effect on the physical, environmental, social and ethical structures of cities in India.

Millions of people live in adequate conditions—without piped water, electricity, security of land tenure, access to roads or health facilities. The means available for production and financing of housing and urban infrastructure are too limited to meet basic needs.

Reducing Poverty and Creating Jobs

Urban poverty is rising at an alarming pace, especially among women. The informal economic sector—which makes a substantial contribution to the delivery of services, production of goods, building of infrastructure and housing construction—often provides the only opportunity for the urban poor to make a living.

Local informal housing construction, for example, generates up to 20 per cent more jobs than high-cost construction. Street hawking, waste recycling and food production are primary sources of income among the urban poor and are illustrative of the creativity of survival strategies.

However, the informal sector itself is often highly exploitative and fails to raise people's economic development beyond mere subsistence. Larger economic strategies and more participatory urban planning approaches that take stock of local skills, technologies and materials are required to generate new and better-paying job opportunities in cities and towns.

Incorporating Environmental Concerns

In 1992, the Rio Conference on Environment and Development designed the Agenda 21 Programme of Action to help save a planet endangered by environmental neglect and plagued by poverty and underdevelopment. Most of the goals agreed to in Rio can become reality only through local action in cities where environmental threats are increasing. Again, it is the urban poor who are particularly endangered by environmental degradation and pollution. The world's Agenda 21 will fail if the city's environmental agenda (population, inadequate sanitation, water supply and waste management) is not addressed. This is being recognized by local authorities all over the world.

Sustainable development in the twenty-first century will to a large degree, depend upon how cities, towns and villages everywhere interact with the environment and utilize natural resources.

Increasing Awareness of Gender issues

Women and men use and experience cities differently, according to their roles, responsibilities and access to resources. For example, when basic services are lacking in a settlement, more often than not it is women who take on responsibilities such as water collection and refuse disposal. Women often have unequal access to resources such as property, credit, training and technology. All of these factors must be addressed urgently, as they make it harder for women to improve their living standards and those of their children.

Disaster Mitigation Relief and Reconstruction

As cities become large and more densely populated, they become increasingly vulnerable to natural and man-made disasters such as earthquake, floods, industrial hazards, epidemics, civil strife and wars. Poor people are forced to live in the most exposed, dangerous and cramped conditions; in flood-prone areas, on steep hillsides or near polluted streams and waste dumps. As a result, they are most likely to lose their homes or their lives when disasters occur. Better planning, access to affordable urban land, and improved construction methods can reduce the extent of catastrophes.

These successful and sustainable approaches to poverty eradication; managing the urban environment; providing access to land, shelter and finance; empowering women and men; and many other issues will have to be documented and disseminated widely.

11

FRESH WATER AND THE ENVIRONMENT

It is widely recognized that water is going to be one of the major issues confronting humanity at the turn of the century and beyond. We are facing a crisis as regards the quantity and quality of water supply, but we have yet to experience the full social and political impact of that crisis. The escalation in the population and the quest for continued development is leading to conflicting pressures on water resources. Such resources are the ultimate recipient of pollution from various socio-economic activities associated with urbanisation, agriculture, mining and clearing of native vegetation. Pollution originating from human waste, especially where appropriate sanitation facilities are not available, or are located too close to water supply sources affects both surface water and ground water.

This makes water supply and health perhaps the most important issue for the large proportion of the global population. Paradoxically, the demands for "sustainable management" and increasing global population require more potable water from a declining available potable water base.

It is universally accepted that proper water administration is a critical component of sustainable development—that is, development that meets the needs of both present and future generations. Indeed, water is an essential factor in a large number of productive activities, of which one of the most important is the production of food by irrigation. This activity, accounts for two-thirds of the water resources used by humanity. A supply of drinking water and sanitation in urban centres are crucial for preserving human health.

For some decades it has been known that the misuse of water resources is responsible for many important environment problems. For example, in many industrialized cities both surface water and ground water are seriously contaminated. This deterioration is a consequence of a range of human activities, sometimes in isolation, others over a large area or a long period of time. Among examples of the latter is modern agriculture, whether it uses irrigation or not, as a result of the intensive use made of mineral fertilizers and pesticides.

Water Shortage: Exaggeration, Reality or Bad Management?

Some of these problems have made news and have created the impression that water shortage will be one of humanity's big problems in the coming decades. Sometimes this feeling is due to genuinely manipulative publicity campaigns to justify the setting in motion of hydraulic megaprojects which basically benefit a few large construction companies. The truth is that except for a handful of very specific cases, no problems of water shortage are to be found almost anywhere. On the other hand, cases of bad water management are not rare at all.

Basic Principles for Good Water Management

Good management of water resources—and of almost all other natural resources—must be based on the principles of solidarity, "subsidiarity" and participation. The physical reality requires that these resources be considered a common heritage of humanity both now and in the future. By "subsidiarity" we mean that water management should be as decentralized as possible: what one person or any minor social group can do should not be done by a higher authority. For example, what local government can do should not be done by regional, state or central government. Participation consists in water users playing as large a part as possible in decisions affecting water, in keeping with each state's or country's social and cultural structure. Obviously this participation calls for a certain cultural and technical knowledge—a hydrological education—on the part of those users.

The need for participation by users is even greater in the exploitation of groundwater. In this case, users tend to extract water independently of one another. They often fail to realize, until there is a serious economic or environmental impact, that their pumping affects other people who rely on the same water supply as has happened.

Water shortage is rarely a serious problem: in fact, in some cases the problem is exaggerated to justify the construction of large works using taxpayer's money. On the other hand, the contamination of surface and groundwater tends to be a problem which rarely receives adequate treatment. Successful water management should be based on three basic principles: solidarity, subsidiarity and participation. The specific way in which these principles are applied will vary from one state or country to another, but the effectiveness of water management will depend in large measure on the hydrological education of the general public.

The universal way of obtaining freshwater is from rain. River systems are the results of the excess water that falls on dry land in the form of rain. On the one hand, rainwater penetrates the permeable soils, saturates them and accumulates to form groundwater reservoirs, or aquifers, which can come to the surface in the form of springs. On the other hand, the water is absorbed by vegetation, which uses it for pumping minerals and then evaporates it by transpiration. Some rainwater is lost because it evaporates immediately on falling on impermeable surfaces like the asphalt of roads and cities. Running water courses finally flow over saturated soils, shaping the complex systems of the watersheds or river basins.

Since each basin's natural system has developed gradually and has grown up according to the yearly distribution and fluctuations of rainfall, we have to appreciate that any large-scale project for redistributing water by means of pipes, as if it were gas or electricity, is a journey into the unknown. This is because it destroys the results of the work of shaping the climate, however transitory it might be.

Variable Volumes

All water supplies are of variable volume. Both the discharge of rivers and the level of lakes and aquifers depend on rainfall. As these resources are components of a large system, the river basin, a reasonable policy would be to manage water resources according to the characteristics of each basin. This would require, first of all, a proper understanding of the system so as to adapt use and consumption to the existing supply. Conserving river systems as much as possible in their natural state is the best guarantee for the preservation of the landscape and of a constant supply. Groundwater reservoirs aren't canals, but are more like lakes, so that pollution leads to the build-up of a debt which is paid in years to come.

Consumption

Water consumption has increased in recent years as a result of not only population growth but also an increase in living standards. In the rural areas the introduction of new farming methods, the spread of irrigation and the excessive use of fertilizers and pesticides causes very high consumption—it is estimated that more than 2/3 of water consumption is used in irrigation. Agricultural pollution also endangers both surface water and aquifers, which receive water full of chemical products. Many cases of eutrophication, the enrichment of water by nutrients that accelerate the growth of algae, derive from the run-off of fertilizers. The practice of intensive stock-raising on farms with large numbers of animals also brings about these problems of over consumption and pollution. Cleaning the stockyards requires large amounts of water which is then released into the environment with high concentrations of nitrogen.

As for industries, they have in the past taken little care over water consumption and dumping, and in many areas the need for proper attention comes as something new. The best thing would be to make industry take its water at a point down-river from where it returns it or, better still, generalise the use of closed circuit systems based on the constant recycling and reusing of the same water.

As regards human consumption, the general attitude to cleanliness is based on diluting pollutants. One example is the success of the use of the Water Closet which involves diluting a few decilitre of urine in 10 or more litres of drinking water: quite a record in wastefulness.

Another aspect to be considered is the different quality of the water that falls on well formed soils from the water that falls on roads, cities, airports, suburbs and built-up areas and whose composition is less stable and "worse" than that resulting from a more uniform interaction with mature soils. Remember that streets, roofs, communication routes, airports and built-up areas already cover a high proportion of the earth's land area and are still on the increase.

Purification techniques should be based especially on the natural processes that include biological activity. Otherwise—for example, if physico-chemical methods are used—there can be side-effects such as an excess of mud or sediments. The strategy to follow is to optimize operations in our use of water according to the discharge and to the distribution of contamination. A system in the form of a conduit or channel, such as a river, can respond relatively quickly. On the other hand, lakes and dams can only do so up to a point, because they show more inertia and irreversibility and take longer to clean.

Large lakes, not to mention the sea, might seem a good place to dump contaminating refuse, but they can't then be cleaned. This is the price we pass on the future generations: a comfortable attitude, but an unacceptable one.

12

FORESTS

Global losses of forest area have marched in step with population growth for human history. The two trends rose slowly for millennia, turned upward in recent centuries, and accelerated sharply after 1900. Indeed, 75 per cent of the historical growth in global population and an estimated 75 per cent of the loss in global forested area have occurred in the twentieth century. The correlation makes sense, given the additional need for farmland, pastureland, and forest products as human numbers expand. But since 1950, the advent of mass consumption of forest products has quickened the pace of deforestation.

In some cases, population pressure is still closely linked with deforestation. In Latin America, for example, ranching is the single largest cause of deforestation. Because most meat produced in Latin America is consumed there, and because meat consumption per person has been largely unchanged for several decades, it is likely that expanding population is the principal reason for ranching-related deforestation. In addition, analysts at the World Resources Institute estimate that overgrazing and overcollection of firewood—which are often a function of a growing population—are degrading some 14 per cent of the world's threatened frontier forests (large areas of virgin forest). In fact, a U.N. Food and Agriculture Organisation study showed a one-to-one correlation between population growth and fuelwood consumption in 16 Asian countries between 1961 and 1994.

On the other hand, deforestation created by the demand for forest products tracks more closely with rising per capita consumption in recent decades. Global use of paper and paperboard per person, for example, has doubled (or nearly tripled) since 1961, and most of the increase has come in wealthy countries with low or even stable levels of population growth. Europe, Japan, and North America, with 16 per cent of global population, consume 63 per cent of the world's paper and paperboard and nearly half its industrial wood.

Although consumption and population growth have operated somewhat independently in the late twentieth century, the two forces could coincide in the developing world in coming decades, with substantial consequences for forests. Developing country paper consumption is less than one tenth the level found in industrial nations, suggesting that large increases in consumption are likely as these nations prosper. (It also suggests that greater economy is needed in industrial countries). With 80 per cent of the world's people, and as home to all the increase in population in coming decades, even modest growth in per capita paper and wood consumption in developing countries could place substantial pressure on forests. If paper were used by the entire world in 2050 at today's industrial-nation rates, paper production would need to jump more than eightfold over 1996 levels.

This projected growth is unsustainable, given that global use of forest products is already near or beyond the limits of sustainable use. Using data on sustainable forest yields, and assuming that virgin forests are left intact, researchers at Friends of the Earth UK have determined that production of forest products for the world is 25 per cent beyond the most restrictive estimates for sustainable consumption. (Many forests, of course, are already logged well beyond sustainable levels). The most optimistic assessment would allow for a further 35-per cent growth in consumption. Even that spells trouble, however, given a projected global population increase of some 54 per cent over the next half-century, and given the likely increase in consumption from rising prosperity. Lower consumption of forest products and increased recycling in industrial countries can make room for a more prosperous developing world to enjoy the products of the world's forests, but the task will be made easier if population growth everywhere is stabilized sooner rather than later.

If population and consumption eat into the world's forests, the resulting loss of forest services reduces, in turn, a country's capacity to support its population. Forests provide habitat to a diverse selection of wildlife; tropical forests, for example, are home to more than 50 per cent of the world's species. And as storehouses of carbon, forests are key to regulating climate. Deforestation leads to huge releases of carbon: an estimated one quarter of the world's carbon emissions come from forest clearing. Loss of these macroservices undermines the stability and resiliency of the global environment on which economies—and populations—depend. In addition, forests provide services vital to a local population, such as control of erosion, steady provision of water across rainy and dry seasons, and regulation of rainfall. Taken together, the loss of these services due to deforestation can upset local economies and subject local populations to economic instability.

13

TOWARDS HEALTHY CITIES

More than a third of the urban population in developing world live in housing of such poor quality with such inadequate provision for water, sanitation drainage, garbage collection and health care that their health is constantly under threat. But, properly planned, cities can be safe and healthy.

In the cities of India, it is common for one child in three to die before the age of five and for virtually all infants, children and adults who survive to have disease burdens many times higher than they should.

Diarrhea, tuberculosis and respiratory infections (each among the largest causes of death) are generally much increased by over-crowding. Many accidental injuries happen when there are three or more persons living in each small room in shelters made of flammable materials and there is little chance of providing occupants (especially children) with protection from open fires or stoves.

But cities also include some of the India's safest and most healthy neighbourhoods. High densities allow much lower costs for supplying each household with piped, treated water supplies and most forms of health, educational an emergency services.

Sanitation and drainage may be costly in cities, as complex systems are needed to cope with high densities and large population concentrations but city households can generally afford to pay more—and are prepared to do so if they get a good service.

Cities may be considered ecologically unsustainable because of high consumption and waste levels but well-planned and managed cities can combine high living standards with remarkably low levels of energy consumption, resource-use and wastes. The concentration of people and production creates many more possibilities of collecting and recycling wastes and for walking, bicycling and a high quality public transport.

For many, city-life is one of excessive workloads and drudgery, yet cities remain centre of culture, including the visual and decorative arts, music, dance, theatre and literature. Most cities have a large reserve of young people on whose initiative and energy they could draw to improve condition, yet most such people find that their cities offer them little hope and little prospect of employment. If cities have such potential to provide healthy, stimulating and valued places to live and work for all age groups, why do so achieve this?

Supporting Change

Much of the explanation is the lack of 'good governance'. Good governance in any city means encouragement and support from all levels of government for a great range of investments of capital, expertise and time by individuals, households, communities, voluntary organisations and NGOs, as well as private enterprises. In most cities in India, the total value of investments made by people in their own homes and neigbourhoods exceeds many times the total value of capital investments made by municipal authorities. Yet governments and aid agencies usually ignore (or deem illegal) most such efforts.

Most households who want their own home cannot afford to purchase one, or at least one that is legal. They cannot obtain housing loans so the cost of the house purchase can be spreads over a number of years, as they cannot meet the (usually) inappropriate conditions set by banks or housing finance institutions. If they turn to building their own home, as most do, they have to occupy or purchase the site illegally. They often have to build on dangerous sites—in floodplains or on slopes with frequent landslides or mudslides—as the cost of safer sites is too high.

Even if they can qualify, for a housing loan; most such loans are for finished houses, not for incremental construction. And even when they have developed their own home and neighbourhood into a viable residential area, governments usually refuse to provide these with roads, water supplies, drains and other essential infrastructure, because they are 'illegal'

What would cities look like today if governments had supported these individual and community efforts by ensuring that land, building materials, credit and technical advice were as cheap and readily available as possible? Or if government-community partnerships had been formed to, at least, improve water supply, sanitation, drainage and health care.

These work within what is often called the 'social economy'—the great variety of initiatives and actions that are organized and controlled locally and that are not profit-oriented. The social economy includes the work of citizen groups, residents' associations, street or barrio clubs, youth clubs, and parent associations that support local schools. It includes many voluntary groups that provide services for the elderly, the physically disabled or other individuals in need of social. It often includes many initiatives that make cities safer and more fun-helping provide supervised play space, sport and recreational opportunities for children and youth. It may provide formal or informal supervision or maintenance of parks, squares, and other public spaces.

The social economy not only 'gets things done' but also creates a dense fabric of relationships that allows citizens to work together in identifying and acting on local problems. Its value to a 'healthy city' is enormous, even if it is often forgotten by governments and international agencies.

The capacity of city authorities to govern is not the same as the capacity to invest, since these authorities can do much to encourage and support the social economy. City authorities can often greatly increase the supply and reduce the cost of land for housing by changing inappropriate regulations, streamlining planning and land-use control, procedures and making better use of publicly owned land.

City authorities should also have the main role in enforcing legislation on, air and water pollution and occupational health and safety. This does not require large investment by public authorities, but it can do much to improve health and the quality of life in a city. Good governance also means managing competing claims and finding common ground between enterprises, trade unions and residents about what should be done to make the city more healthy.

Achieving a healthy city needs a representative political system through which the priorities of citizens and businesses can influence policies and actions. Democratic structures remain among the best checks on the misallocation of resources by city and municipal governments. Actively involving a wide range of local groups in developing 'city governance' helps ensure that the different priorities of a wide range of groups are addressed.

The key issue is not so much identifying what should be done to achieve more healthy cities. This is well-known. It is identifying how it should be done, especially how governments and international agencies can support a vast range of activities by individuals, households and communities that help build and maintain healthy cities—which to date they have ignored or even (for many governments) repressed.

GLOBAL WARMING:
Worrisome Signs

Scientists increasingly agree that the earth's atmosphere is becoming warmer. A long-term rise in the global climate could cause sea levels to rise around the world and bring a number of other adverse consequences. Reliance on fossil fuels as an energy source and the widespread destruction and burning of forests are chiefly responsible for the carbon emissions—the so-called greenhouse gases—that lie behind global warming.

One indication of global warming is that over the past 40 years the ocean surface (the top 1,000 feet) have warmed an average of half a degree Celsius. The US National Oceanic and Atmospheric Administration (NOAA) has reported that tropical waters in the Northern Hemisphere have been warming up even faster—in fact, 10 times faster than the measured global rate—because tropical oceans retain heat more readily than other areas.

Rising Sea Levels

Studies project that by 2100 the earth's surface temperature could increase between 1.0 and 3.5 degrees Celsius. If the highest projection were reached, Greenland's ice sheet probably would melt. As a consequence, the global sea level gradually would rise as much as seven meters.

Computer models project that this rise in sea level would take more than a millennium. Some climatologists, however, think that sea levels could rise much faster, pointing to dramatic shrinkage of the Arctic ice cap over the past 30 years.

Even a rise of one meter in sea level—which could occur by 2080, according to the computer models—would inundate many low-lying coastal areas around the world. For instance, much of the Nile River Delta of Egypt would disappear. A one-meter rise in global sea levels also would inundate close to 20 per cent of the coastline of Bangladesh and displace millions of people.

Adverse Health Effects

Rising global temperatures also would carry adverse health consequences. As temperatures warmed and episodes of droughts and floods became more frequent, the incidence of water-borne diseases and a resurgence and spread of infectious diseases carried by mosquitoes and other disease vectors probably would increase.

Warmer global temperatures also would magnify the effects of human activities on the environment, including more pollution and habitat destruction. Climate change might even cause some ecosystems to exceed critical thresholds, leading to their irreversible decline.

Growing Scientific Consensus

In 1988, to help study and focus attention on the issues, the Intergovernmental panel on climate change was created under the auspices of the World Meteorological Organisation and the United Nations Environment programme. The panel has involved as many as 2000 scientists from around the world. In 1996 a panel report concluded firmly that global climate change is a reality and not just a possibility. After reviewing the evidence, the panel determined that:

- Evidence for the link between climate change and human activities is compelling. Already, increase in carbon dioxide and other climate-changing gases have upset the balance of the earth and its atmosphere.
- The earth's surface has become warmer; the number and severity of storms have increased; and the global sea level has risen by 10-25 cm. over the past century.

Because the warming trend is a global problem, solutions must be global in scope, the panel concluded.

Why is the Climate Changing

Over the last 150 years burning of fossil fuels has released some 270 billion tons of carbon into the atmosphere in the form of heat-trapping carbon dioxide gases. Since 1950 annual worldwide carbon emission has increased fourfold, reaching 6.3 billion tons in 1997. Other emissions that contribute to climate change include methane (mainly from domestic livestock and agriculture), nitrous oxide, and chlorofluorocarbons.

Atmospheric concentrations of carbon dioxide reached 363 parts per million in 1998, the highest level since the time of massive Volcanic activity over 160,000 years ago, based on examination of ice cores in Antarctica and in the Arctic. If current trends continue, atmospheric concentrations of carbon dioxide would double during this century.

About three-fourths of the huge increase in carbon emission over the past half-century is due to increased energy consumption per capita; about one-quarter is due to population growth. Western industrialized countries account for nearly half of atmospheric carbon emission, but developing countries are producing a growing share as industrial activity increases and populations grow. China is now the world's second largest carbon emitter, after the US.

Vanishing Carbon Sinks

The earth's forests are carbon sinks that currently soak up an estimated one-third of the carbon dioxide released into the atmosphere. When forests burn, whether naturally or when people clear the land, they not only release more carbon into the atmosphere but also diminish the amount of carbon-absorbing forest cover remaining.

Some scientists are concerned that droughts caused by global warming will increase the number of forest fires, thus contributing further to carbon emission in the atmosphere. For instance, the six months of extensive forest fires that occurred in Asia in 1997 and 1998 released more carbon into the atmosphere than western Europe emits in a year. Burning trees for land clearance in the tropics releases about 1 billion tons of carbon into the atmosphere annually.

As more carbon fills the atmosphere, scientists worry that forests will become saturated and no longer play their role as carbon sinks. Instead, they will start to release carbon themselves.

Agriculture at Risk

Higher carbon dioxide levels in the atmosphere would extend the agricultural growing season and promote forest growth in the short-run but would have potentially negative effects on crops and forests in the long-run. Because the world's grain belts would become less productive, an additional 350 million people would go hungry by the middle of this century. Major droughts have been projected for sub-Saharan Africa as climatic patterns shift, reducing rainfall and drying out soils for longer periods.

In 1999 NOAA projected that by the middle of this century soils in agricultural regions of the central US, Central Asia, and the areas surrounding the Mediterranean Sea would likely experience substantial reductions in soil moisture during the summer growing season because evaporation rates would be higher. Such reductions in soil moisture would make these areas particularly vulnerable.

Others point out that, ironically, global warming could produce colder temperatures in Northern Europe and Russia, reducing crop yields in these regions as well. This change would occur because the huge amounts of arctic freshwater from melting ice caps would make the water less dense. Such a change would interrupt the "conveyor belt" effect of the North Atlantic Drift, the ocean current that transports warm tropical water from the Gulf Stream to Scandinavia and Northern Europe.

What Can be Done?

What is the prospect for reducing emissions of carbon dioxide into the atmosphere? The United Nations Framework

Convention on Climate change was opened for signature at the Rio Earth Summit in 1992. It was promptly signed and ratified by most low-lying island states and countries with extensive coastal areas. The Convention established a framework and a process for agreeing on specific actions later on; it asked signatory states to take preliminary action to reduce greenhouse gas emissions; and it encouraged scientific research on climate change.

15

MONEY ALONE IS NOT ENOUGH:
Technology Transfer and Environmental Protection

In the seventies it was a hotly debated topic, in the eighties it became a moot issue: The demand of the developing countries for low-cost or even free technology transfers from the industrial nations. The environment, or more accurately, the endangered environment, is responsible for reviving this subject once believed to be dead. Politicians in the South were quick to see the opportunity which presented itself: No environmental protection without technology, no technology without technology transfer, not technology transfer without money.

The Montreal treaty (on the reduction of chlorofluorocarbon production) was an important first step. It established a fund which supports the environmental efforts of the developing countries. But this was only the beginning. Technology was one of the main concerns of the United Nations Conference on Environment and Development (UNCED).

It is undisputed that private enterprises control the expertise necessary for environmentally sound technologies. The discussions in the developing countries revolve around this basic issue: What guarantees are there that these firms will transfer any technologies at all and at an acceptable price to boot? Both premises present a problem: In all likelihood, technology monopolists have invested substantial amounts in the development of the respective technology and will therefore try to sell their licenses at the highest possible price

(price problem). Having had so many failures with specific projects in the past, many firms are not quite reluctant to transfer technology to the developing countries. It was no coincidence that north-South technology transfers practically came to a standstill in the eighties.

Technology cannot be purchased as a package. This is a frequently forgotten truism. By general definition, technology consists of four components:

- Hardware, for instance a specific configuration of machines and equipment to manufacture a product or provide a service.
- Know-how, i.e. scientific and technical knowledge, qualifications, and empirical knowledge.
- Organisation, i.e. the arrangement which combines hardware, know-how, and operational management methods.
- The end product, i.e. the item or service.

The technical components (machines, blueprints) and products (licenses) can be purchased subject to the cited restrictions, but not organisation and qualifications. This is the real bottleneck for technological development in most developing countries. There is no inducive to the process which is most important for the utilisation of technological learning.

The core of technological knowledge is the mastery and subsequent continuous improvement of production processes. In part this happens automatically (learning by doing), but beyond that it must be actively stimulated. Many examples in both industrial and developing countries show that firms frequently stagnate at a certain technological level, thus missing a chance to improve efficiency. The main reason is inadequate technological knowledge. The secret of optimizing the conversion process lies in a strategy of small steps, the steady improvement of individual segments.

The economically most dynamic developing countries are successful because of high productivity increases made possible by technological competence, i.e. the ability to assess and evaluate the technology offer, to select, utilize, and improve technologies, and ultimately develop new ones. This latest state-of art processes can be used for industrial expansion projects. As far as these countries are concerned, the introduction of financing mechanisms for the transfer of environmentally sound technologies represents a very promising approach.

Technological Competence—the Crucial Factor

In other countries industrialisation efforts have caused serious environmental degradation but little economic development. The reason is last but not least inadequate, only slowly growing technological competence. Those countries have hardly any money to invest in environmentally sound technologies. But even if the international community establishes financing mechanism, the problem of inadequate technological competence remains unsolved. It is unreasonable to assume that in a country, where conventional production technologies are used ineffectively and inefficiently, environmentally sound technologies can suddenly be applied in a meaningful and efficient manner. For these countries the transfer of environmentally sound technologies is a "quick fix" which in all likelihood will not work. Without an established level of national technological competence, it will not do much good to shower a country with technology from outside (more precisely, with technical hardware and production know-how).

This brings us to an original development policy problem. Technology and technology transfer have always played a big role in development policy, although quite often the perspective was short-term: Instead of technology, only hardware was transferred; frequently, technological competence was not developed in the recipient countries, but rather substituted with external experts. In the future, much more emphasis will have to be placed on stimulating the technological learning process and promoting national technological competence. Many developing countries have already moved in this direction, such as instituting macro-political reforms, which pressure private industry to increase performance, thus forcing technological learning. But development policy can make a contribution as well. Above all, it will have to adopt a more systemic approach and promote structural improvements at several levels:

- As a systemic link between project and project environment. Technology institutions have always been the darling of environmental policy, but frequently they had too little contact with potential users and therefore remained ineffective.
- As linkage with indigenous efforts in the recipient countries and with the activities of other donors. Isolated projects and competition among donors are a guarantee for failure. On the other hand, the fascination in many recipient countries with individual technologies is slowly being replaced by a growing understanding of the technological correlations, concerted technological-political actions, which have already been initiated in some developing countries (for instance Thailand, Jordan and Tanzania) and which brought together representatives of government, private industry, educational and research institutions at "round table" discussions, provide an opportunity for the systemic incorporation of technologically-oriented development policy measures:

A Dual Challenge for the Industrial Nations

The industrial nations are thus faced with a dual challenge. First, they must support those developing countries financially whose technological competence is adequate for the effective utilisation of environmentally sound technologies. On a global scale, this is an ecologically rewarding undertaking. Since environmental standards have been low in these countries, investments can achieve substantially higher reductions of pollutants than in the industrialized North. There is another aspect: The industrial nations can put pressure on the developing countries to use environmentally sound processes only if they simultaneously offer financial compensation. Secondly, the industrial nations must increase their efforts to raise the level of technological competence in developing countries. This is an essential pre-condition for the developing countries to be able to participate in the medium-term in an environmentally benign growth model.

16

TOURISM AND THE ENVIRONMENT

The relationship between tourism and the environment is obvious, and is largely established through what is sometimes called "environment quality". This quality is perceived in different ways according to the human population and the circumstances presiding tourist activities at any given moment. Any analysis of the relationship between tourism and the environment that we can include under human ecology therefore comprises aspects of the natural sciences as well as the social sciences.

Tourist activity is promoted, conditioned and influenced by the environmental circumstances of each region and can be affected by modifications or changes in those circumstances. Although a lot of emphasis has been placed on the negative impact or modifications in "environment quality" attributed to tourism, it is also accepted that it can be a very important factor in the preservation and defence of ecological values threatened by more destructive alternatives for the use of territory. Very often, tourism can be the most suitable and most satisfactory way of using a region's renewable natural resources. Nevertheless, their management and use need to be properly regulated so as to guarantee their renewability and persistence.

There is room in this complex field of relations to study, rationalize and optimize an activity as important as tourism, from the point of view of its insertion in the ecological systems with which it interacts. However, there are relatively few efficient studies on issues of real importance. It is startling to observe that places with tourist potential undertake little or no research in this field.

One possible cause is the difficulty in identifying the real problematic in tourism/environment relations, which is essentially interdisciplinary and involves the integration of traditionally separate areas of knowledge. Although work is undertaken from time to time on environmental psychology, the sociology of tourism, behaviour in relation to the environment, etc., they are very rarely combined with works on the environment, forestry and agricultural policies, soil use, contamination, biodiversity, evaluation of environmental impact, nature conservation, etc., in search for a more integrated management of tourist resources.

Responsible Tourism

Tourism runs the risk of going the way of other phenomena, which first of all experience rapid growth and then suffer a spectacular collapse, what in economics is often called "boom and bust"

The causes are familiar: a certain dose of greed, often based on a lack of mid-or long-term planning, property speculation, little consideration for local populations—in both economic and social aspects—and, in general, a lack of awareness as regards environmental aspects—contamination, water use, energy, etc., -on the part of tour operators, hoteliers and other agents involved in tourism in its different forms, including the tourists themselves. The problem is particularly evident in ecotourism, based on the wonders of the natural world: landscapes, flora and fauna. Many experts fear for the future of this type of tourism, which has grown spectacularly in the last few years. Landscapes deteriorate, the fauna decreases, the designers and administrators of tourist developments fail to respect the most elementary principles for adapting architecture to its surroundings, or else there is little effort to recycle, economise or educate with a few honourable exceptions, tourist planning is careless and irresponsible.

And yet a responsible approach would be in the tour operators' own interests, as it would make the tourist industry sustainable, with positive influences on biological, economic and social aspects.

Ecotourism, for example, has shown that when properly conceived it can become a powerful instrument for the preservation of nature, with very favourable repercussions for local populations and for educational programmes, while offering hundreds of millions of ecotourists a wide range of spiritual and physical satisfactions. At the same time, the host countries can take pride in what they have to offer their citizens and the rest of the world.

The preventive and corrective measures are known to us; what is needed is a sense of responsibility and farsightedness on the part both of the authorities and of the industry. We need regulations and controls, so as to put the people who do the damage out of circulation an reward those at the forefront of sustainability

Sustainable Tourism

After several decades of rapid quantitative growth, tourism is going through a period of profound transformation. Tourists, the consumer in this industry, but also the public, have started to demand a change in the conditions of production and use of tourist service, putting an end to the uncontrolled expansion of mass tourism.

This is the ultimate reason, apart from ethical and aesthetic considerations, why tourist activity as a whole, in the private sector as well as in the public and voluntary (NGO) sector, has begun to seriously analyse the implications of tourism in terms of socio-cultural and environmental impacts, and to consider the need to draw up and implement environment-friendly tourist policies.

Indeed, while not denying the viability and the utility of alternative approaches of an external and coercive nature, it is obvious that the decision-makers in the sector react better to positive stimuli. The realisation that their clients prefer well-conserved areas and non-aggressive tourist practices and that they are prepared to pay more for this makes it easier to adopt strategies of sustainability in the tourist industry in a sincere alliance with conservation movements.

All this points to the validity of Overall Quality Management as a viable method in sustainable tourist activities. The overall quality approach renders the management of products and especially of tourist areas extremely sensitive to the preferences and expectations of consumers. The private public profitability of a tourist destination will depend on client's satisfaction, since these will return more often and for longer and will pass on a positive image of their holiday experiences. In so far as these preferences and expectations include the demand for unspoilt settings, consumer satisfaction, and therefore the profitability of a tourist spot, will call for the development of strategies for sustainable development.

One can believe this is a productive approach for sustainability in the tourist business and one that makes for professional attitudes that fit in with the economic target of businesses and other organisations. There is only one prior requirement: continued education and training of everyone involved in tourism, from consumers to those responsible for tourist policies. The demand for quality, and even more so for environmental quality, is a call to people's awareness, to their understanding of the environmental and cultural implications of any activity and their ability to express themselves and to organise to choose the most clear-sighted line of action.

Tourism in the Modern Age

What will the tourist trade of the year 2000 be like? Who will be the tourists of the coming millennium? These are the questions which, faced with the extraordinary boom in tourism, expert, tour operators and politicians have repeatedly posed over the last fifteen years. These questions arise either because of the financial profits the tourist industry involves, or from the demands of consumers who show new awarenesses, habits and lifestyles. In the eighties, mass tourism gradually changed and people began to talk of "tourisms". Expressions such as cultural tourism, sports tourism, religious tourism, adventure tourism or ecotourism have become part of everyday language. In the past, the dominant practices was to take one long holiday in a single destination. Today, people tend to distribute their holidays over different destinations and different times of the year.

From a socio-historical point of view, three types of tourist industry can be differentiated. In the case of the industrial tourist, for whom work is the center of existence, the motivations for travelling can be summed up as rest and freedom from responsibilities. This type is gradually decreasing in number. The hedonistic tourist belongs to the generation that discovered entertainment and consumerism. They like to go on holiday to experiment, to explore the unknown, enjoy themselves meet other people and relax in unspoilt natural surroundings. These are the majority today and will continue to be so. Finally, the modern age tourist, someone who tends to reduce the polarity between work and play: not just work, but just not fun, either. Their reasons for travelling include broadening their personal horizons and getting back to simple things and nature, with a touch of creativity in the planning of their journey. These are gradually growing in number and in future will form an important segment of demand.

One characteristic in the expectations of the modern age tourist is the capacity to make a critical appraisal of the offer and to influence it. Producers should be more attentive and sensitive to the new demands and be flexible enough to cater to the tourist in search of higher quality. In the third millennium in fact, the concept of quality will have to take environmental aspects more into account. Recent forms of tourism point to a renewed interest in nature and a wish for quality tourism. So much so, that some tourist spots are reorganising their own offer in keeping with these trends. Quality is the result of a complex strategy which is organised day by day. The consumers, whose environmental awareness is constantly growing, will expect to identify, verify and be able to differentiate ecologically correct products from the imitations now invading the market.

The present millennium is coming to an end and is leaving Western countries with a high level of welfare and a large tourist demand to satisfy. Nevertheless, serious environmental problems also plague areas that receive a high influx of tourists. Tourists, tour operators, local authorities and the general public are therefore called on to find new forms of coexistence and the right solutions for themselves and for the survival of the planet.

17

ENERGY AND SUSTAINABILITY

Mankind's history is marked by a growing use of energy which until the end of the Industrial Revolution came largely from renewable sources. It was coal that fed the furnaces and boilers of the Industrial Revolution from the end of the seventeenth century to the nineteenth century, and drove railway transport and steamships. As well as being a useful source of mechanical energy, it was also used in the manufacture of coal gas for street lighting and in the chemical industry. In fact, coal was the principal form of energy until 1900.

Discoveries at the beginning of the nineteenth century allowed the use of electricity and revealed the relations and the interconvertibility of different forms of energy. The principles of conservation and of energy quality did not become operative until much later. Meanwhile, in 1882, the first system for producing and distributing electricity in a large city was installed. This was the beginning of the second phase in industrialisation through electrification.

Following the first successful oil drillings in 1859, Standard Oil, the first of the modern large scale oil companies, attempted the first vertical structure for overall control of the oil process. It involved extraction from the subsoil, storage, refining and final distribution. Later, the growth of derivatives, the lower extraction costs compared to coal and the greater ease and economy of transport made oil modern society's basic energy source.

The internal combustion engine led to motorisation on a massive scale by land, sea and air and guaranteed a constantly growing market for petrol. The forties marked the start of the new petrochemical industry, which gave rise to an enormous number of new products; synthetic rubber, plastic, medicines, cosmetics, varnishes, artificial fibres, detergents, weedkillers, fertilizers, butane, propane, etc., opening the way to the mass-production of consumer goods and introducing new, non-biodegradable substances into the environment.

After World War II, ambitious programmes to produce electricity from nuclear energy were begun, in the search for a return on the enormous amounts of money invested. The economic expansion in the West during the fifties and sixties was directly related to enormous petrol consumption at a time when energy was considered plentiful and cheap. Energy consumption during these decades grew more than exponentially. The fastest developing industrial sectors were precisely the ones that consumed most energy—petrochemical industries, metallurgy, car manufacturing, domestic appliances, electricity generating, etc.—and a trend developed towards goods and services with higher energy intensity. Since 1950, increased energy production has been systematically favoured over more rational use. So much so that the increase in energy consumption has been taken as a reliable indicator of progress.

The Aftermath of the Oil Boom

The oil crises of 1973 and 1980 showed up the fragility of an energy system that was over-dependent on oil. The War in the Gulf was reminder of what was at stake for the Western economies; free access to cheap oil in the Middle East. It was therefore fear of the hardship caused by the first crisis that brought about a change in attitudes in Western countries; efforts were directed at breaking free from this dependence, diversifying supply sources, perfecting replacement energies and promoting energy-saving programmes.

The eighties marked a change in people's awareness about environmental problems. The damage was making itself felt in more and more places and eventually the global threat to our planet as a result of our energy system became clear; the composition of the atmosphere was changing and could lead to possible changes in the climate.

According to recent figures, 82 per cent of all the energy consumed in the world is produced by burning fossil fuels, 7.5 per cent from burning biomass, 5.5 per cent from the use of hydraulic energy and 5 per cent from nuclear energy. Most of our energy in other words, in non-renewable; it runs out as we use it, as the population increases; and it comes from fossil fuels, which on burning increase the amount of CO_2 in the atmosphere. If we add to this the accumulation of nuclear waste, the problems of access to oil deposits, constant spillages during transport and all the different imbalances involved in the world energy system, the outlook is far from sustainable.

The inequalities speak for themselves; globally, less than a quarter of the world's richest population consumes almost three quarters of the energy commercialized in the world. For example, the average annual consumption per capita in the United States is 26 times higher than in India.

The Choice of Change

Opening the way to societies that make sustainable use of energy necessarily involves increasing and improving energy efficiency, both in supply technologies and in end-use technologies, at the same time using renewable energy sources instead of fossil fuels.

Choosing the right system for the transformation of primary energy sources into energy services such as lighting, cooling, cooking, mechanical force, transport, etc. and choosing the most suitable appliances and technologies in each case is fundamental.

The truth is that a good standard of living is possible without wasting anything like as much energy. A series of relatively straightforward measures today allow a far higher level of comfort than in 1950, using one-third as much energy for heating water for washing in the home.

Petrol consumption by vehicles has dropped by 40 per cent in forty years, from 8 litres/100 kilometres to 5.3 litres in some models, and the work of improving their energy efficiency continues. In industry, the energy consumption necessary for manufacturing large intermediary products (steel, cement, paper or fertilizer) is decreasing steadily at a rate which varies between 0.5 per cent and 0.2 per cent per year according to the product.

Today's incandescent bulbs consume one twentieth as much electricity as bulbs in the twenties. The compact fluorescent bulbs now available can cut this down again to one-fifth. Efficiency in lighting has increased one-hundredfold. The use of new materials and a more rational use of traditional materials allows a reduction in the amount of energy and raw materials consumed. Building a house, for example, requires 20 per cent less energy than in 1950; building a vehicle, 40 per cent less. On a global level, reducing our society's energy-intensiveness is the first step towards energy sustainability.

18

CITIES RESIDENTS TO THE RESCUE

In the next ten years, the number of people living in cities will rise to around 3.3 billion. Tokyo already has a population of 27 million, Sao Paulo (Brazil) 16.4 million, and Bombay 15 million. World Bank forecasts show as much as 80 per cent of the developing countries, economic growth occurring in the cities and major conurbations.

There are both positive and negative aspects to these developments. At each stage in the history of urbanisation, environmental conditions in cities were improved dramatically. The process was often slow, but over time, many epidemic diseases have been controlled, the supply of clean water and the removal of wastes have become routine, the risks of fire have been contained, and standards of comfort and cleanliness have risen to unprecedented levels. Cities could not have become as large and as numerous as they are now if environmental conditions had remained unchanged.

In a curious way, the pollution that cities suffer is largely due to their wealth. The rich consume a great deal more energy, water, building materials and other goods than the poor, and thus produce much more waste. This is what is happening; in the cities where rapid industrialisation is taking place—only the rich enjoy the benefits of piped water and refuse collection.

Increasingly Unsanitary Conditions

There is another, often tragic, aspect to this situation. The poorest of the poor are reduced to living in outer-edge shantytowns in extremely unsanitary conditions and, lacking the resources to deal with the problem, the city as a whole has to endure congestion and air and water pollution. Some towns and cities are expanding at a rate of over 7 per cent a year, municipal sanitation departments are no longer able to cope, and it is estimated that as many 30 per cent of the population are without running water.

In many of the world's major cities runaway population growth, an epidemic of Aids and rising social tensions have been combined in the last few years with a steep drop in incomes. The population living on the outer edges of the cities continues to grow apace, hundreds of thousands of people are without running water and 15 per cent of them without sanitation of any sort. Various voluntary bodies and non-governmental organisations have got together, often successfully.

Water and the Environmental, Crisis

One key problem concerns the availability of clean water. Some progress has been achieved as a result of the International Drinking Water Supply and Sanitation Decade, but in 1994 at least 220 million people still lacked a source of drinking water near their homes. In some cases, communities of 500 or more inhabilitants are served by a single tap. In some towns, communal taps function for only a few hours a day, so that people cannot build up sufficient reserves of water for their personal needs if it takes too long to fetch or if the water to be carried long distance.

As there are no proper sanitation measures, the disadvantaged members of the population have to drink dirty water, fish in polluted stream, and eat vegetables that have been grown by the side of refuse tips.

A further major problem arises from the threefold harmful impact of cities on the environment; urban development on agricultural land, the extraction and exhaustion of natural resources, and the dumping of refuse.

Growing pressure on coastal regions, where nearly a billion people now live, is doing serious damage to the marine environment. Development activities pose a threat to nearly half the world's coasts.

Towns originally offered people a place of refuge, of mutual help and culture. According to nineteenth-century town-planning theorists, they should supply all human needs. They were supposed to be the very stuff of civilisation. That was not to be, and therefore whenever the authorities throw their hands, dismayed by the scale of the problems and lacking the political will, money or resources to cope with them, personal initiatives are those most likely to succeed.

19

CHILDREN'S HEALTH AND THE ENVIRONMENT

Children today live in an environment vastly different from that of a few generations ago. Economic development, increased urbanisation and the consequences of war in many countries have added to the traditional environmental hazards, those problems associated with environmental pollution. Thus, while some traditional children' diseases such as diarrhea, malnutrition and infectious diseases persist in many countries, environmentally-related illnesses such as asthma, respiratory illnesses due to environmental tobacco smoke (ETS), as well as mortality and morbidity due to injuries, are increasing. In childhood cancer in some countries and the potential risks of endocrine-disrupting chemicals are among the emerging health threats that need careful vigilance. Children of lower socio-economic status are likely to suffer disproportionately from all these health threats as a consequence of living in highly polluted environments, poor quality housing, lower levels of education, and of restricted access to environmental and health care services.

Children's Vulnerability

The concern for children's vulnerability to environmental health threats is based on several factors. Children receive greater exposures than adults do because they drink more water, eat more food and have higher breathing rates per unit of body weight. Because they are undergoing rapid growth and development, toxicant effects at specific times may have irreversible consequences. For example, if vital connections between nerve cells fail to form during brain development, there is high risk that the resulting neurobehavioral dysfunction will be permanent and irreversible. Also, because most children have more future year of life than adults, they have more time to develop any chronic disease that may be triggered by early environmental exposures.

Public Health Threats

Asthma, injuries, and the effects of environmental tobacco smoke (ETS) are among the most significant public health threats to children. Childhood asthma is increasingly prevalent in allmost all countries. What causes asthma is not known, but several environmental factors, such as indoor air quality (particularly exposure to the house-dust mite) and ETS, have bee linked with the increase in asthma. In addition, outdoor air pollutants such as particulates; sulphur dioxide and ozone can exacerbate asthma symptoms. ETS, especially smoking by the mother, is a known risk factor for asthma. ETS is also known to cause acute and chronic middle ear disease and is associated with sudden infant death syndrome (SIDS).

Potential for Prevention

The variation in asthma and injury rates and the evidence of the role of certain environmental factors underline the potential for prevention. Public policies should seek to avoid preventable childhood diseases by preventing exposures to environmental agents and considering children's characteristics and susceptibilities in the development of environmental health legislation. Promoting citizen awareness and participation in policy-making through education and access to environmental information are important elements in achieving a safe environment for children. In this context, children are not only consumer with rights, but also citizens who can play an active role towards their own protection.

International Awareness

Several international agreements have acknowledged children's vulnerabilities and have committed their signatories to protect children's health from the effects of a deteriorating environment. This year, many countries will address several of the environmental health threats to children through international and national action. It is expected that a large international collaborative initiative will result under the guidance of WHO and other international organisation.

20

ECOTOURISM OR ECOCIDE?

Ecotourism is the fastest growing part of the World travel business, but whether it destroys more than it protects will depend upon how it is put into practice.

For the travel and tourism industry, ecotourism is the fastest growing 'market segment', generally equated with nature tourism. Interpreted merely as a product, however, it may be ecologically based but not ecologically sound, responsible for sustainable.

To incorporate these vital characteristics ecotourism must adhere to three essential principles: The first is, perhaps, the most obvious. As an industry based on the beauty and diversity of nature, it is evident that it should not deplete or degrade those resources and thus prejudice its own future. Ecotourism must, therefore, be ecologically sound, requiring a two-way link between itself and environmental conservation.

To consider nature without recognizing the link with people will, however, compromise sustainability. It is now widely recognized that conservation cannot be divorced from development issues. The second principle is therefore that ecotourism must be responsible, paying regard to local needs and improving local welfare.

However, to be truly sustainable, ecotourism needs to fulfill the ambitions and expectations of all interests. The third principle, then, is to consider not only the interests of tourism enterprises and organisations, but also visitor satisfaction, the needs of tourists.

If, ecotourism embodies these essential principles, symbiotic relationships between the varying interests should follow, with environmental protection resulting both from and in enhanced standards of living for local populations, continued profits for the tourism industry, sustained visitor attraction, and revenue for conservation. An examination of ecotourism across these dimensions, highlights not only its potential but also its problems.

Local Benefits

The high ground claimed by ecotourism, in terms of its contribution to development, is that, in principle, it offers enhanced prospects for local involvement compared with conventional tourism. As well a moral obligation to incorporate the local people in projects that affect them, such incorporation has important developmental implications.

Tourism income may be captured locally through revenue sharing schemes, through entrepreneurship and labour, and through the sale of tourist merchandise. Tourism can also act as a catalyst, and even provide some of the finance, for the improvement of essential services such as clean water, sanitation, electricity supply and transport. It may also provide an incentive for improved education and skills and the potential for participation in decision-making.

Local involvement also makes sense for conserving natural environments. It has been recognized that, as people realize the benefits from ecotourism, support for conservation increases. Ecotourism may also provide the incentive for the survival of a traditional culture. The cultural and the natural are often inextricably linked to form the composite attraction of a particular ecotourism destination. The terracing of the Himalayan foothills, the hot springs at Tatopani, Sikkim are all examples of the fusion of the natural and the cultural.

Greater local involvement makes practical sense for national and local governments, agencies and operators using local labour, expertise and knowledge. Education is a two-way process, improved understanding of local circumstances is likely to increase project efficiency. Building upon local experience and traditions provides a foundation for wise and successful development, and, simultaneously, an ecotourism asset.

Introduction to indigenous uses of natural products also broadens the base of environmental interpretation. Local involvement is not without its problems, however, revenue sharing schemes may neither benefit the most needy, nor those most adversely affected. Beneficiaries may often be passive recipients, rather than active participants. The emphasis must be on participation rather than patronisation, if traditional livelihood are removed they must be replaced with others.

Local participation, however, often consists of employment rather than entrepreneurship, where constraints of costs of entry, language, education and skills operate. Furthermore, the nature of local employment tends to be low skilled, poorly paid and often seasonal. The higher status, better paid jobs, particularly managerial positions, tend to be occupied by outsiders.

Industry Profits

The integrity of the tourism 'product' is vital to the interests of tourism entrepreneurs, and all those who are associated with them, such as tourist boards, government departments, NGO's and international aid agencies. Tourism operators benefit from public support, increased credibility and demand for associated products. Sound environmental practice often makes good business sense.

There are, however, many practical and institutional obstacles to effective ecotourism management, not the least of which will be the problems of veted interests who are more concerned with short-term profits than with the long-term.

Pressure of Numbers

Another dilemma is the sheer problem of numbers. To confine attention, however, to the consideration of small-scale, more easily managed ecotourism projects, involving small specialist groups paying high prices, is to invite ecocide at higher levels.

Rapid growth rates imply inevitable change. Psychological carrying capacity (as well as other types of carrying capacity) will probably be breached and visitor satisfaction compromised. This is especially true when visitors are concentrated in space and time.

However, much a principled definition of ecotourism is advocated, it must be recognized that so-called ecotourists are not an homogeneous group. The spectrum of participants embraces hard-core nature tourists through to casual day visitors. Their behaviour and consequent impact will vary accordingly. It is essential, therefore, to attempt to match numbers and types of ecotourists with destination characteristics.

Paying for Conservation

Willingness-to-pay surveys of ecotourists across the globe show a consistent response of $10 as reasonable visitor's fee. Certain unique sites, or those harbouring more charismatic species, can support higher fees. The potential revenue for conservation is therefore evident, but often not realized.

Where the fee falls below the amount visitors are willing to pay, the capability to contribute more fully towards conservation remains latent. It is also necessary to ensure that a proportion of revenues accrues locally. Percentages of revenues directed towards conservation vary between sites. Often high proportions end up in central treasuries.

The Challenge

The major role players in ecotourism all have a stake in its sustainable development. Their present and future interests are, in many ways, tied to one another. Given the multitude, and diversity, of stakeholders a completely sustainable outcome is, however, likely to remain elusive.

The grand challenge is to reconcile sometimes complementary, but often conflicting interests. The essential dilemma is to balance demands of ever-increasing 'new-tourists' escaping from the confines and pressures of urban life, and reacting against the characteristics of mass tourism with the needs of the environment, the aspirations of tourism organisations, and, most importantly, the basic needs of the local population.

Although a win-win scenario, where all interests gain, is the ideal outcome, there will often be situations where one interest may gain at the expense of another. National Parks, for example, may bring benefits for conservation and for visitors, but the local population is likely to lose out if they are excluded from their traditional activities.

The situation is fraught with discontinuities. A win situation for one interest in a particular place at a specific point in time is likely to be a loss for another. It is necessary, therefore, to recognize conflicts and identify relative costs and benefits. Arriving at the most sustainable outcome is likely to involve trade-offs. It is unlikely to be optimal either environmentally or developmentally, but, in the circumstances, it will be the most feasible and most practical. And, hopefully, ecocide will be circumvented.

21

SUSTAINABLE TOURISM DEVELOPMENT

Tourism has grown into one of the world's major industries and has thus also become an increasingly important, if complex, issue for environmental policy. Unless it is developed in a sustainable manner, we will be unable to achieve key objectives of global environmental policy such as the preservation of biological diversity, the prevention of climate change or the conservation of natural resources.

Tourism itself depends a lot on the existence of unspoilt nature and landscapes, as well as a healthy environment. If nature is plundered, landscapes are destroyed or water, energy and soil resources are over-exploited, the economic basis of tourism is also undermined. The needs of tourism do therefore overlap with those of environmental protection and nature conservation.

On the one hand, for instance, tourists are becoming increasingly environmentally conscious and are looking to get back to nature and enjoy unspoilt environments when on holiday. On the other hand, however, the number of international tourists is growing constantly. The proportion of long-haul journeys is also increasing steadily, especially in the industrialised nations, where travel is now taken for granted as part of people's lifestyles and has become an important factor in social status. The many different types of travel and holidays are covering more and more countries and regions and, as a result, increasing numbers of previously unspoilt natural environments are being opened up to tourism. This applies equally to coastlines, small islands, coral reefs, rock formations and mountain regions.

There is growing recognition of the need for tourism to develop in a sustainable and environmentally friendly manner. Many countries have, for instance, introduced regulations which require environmental impact surveys to be carried out at least for larger tourist developments. Since the Rio Summit in 1992, there have also been more initiatives in support of sustainable tourism at international level.

- Sustainable tourism allows for the rational use of biological diversity and can contribute to the preservation of that diversity.
- The development of tourism must be controlled and carefully managed so that it remains sustainable.
- Particular attention must be paid to tourism in ecologically and culturally sensitive areas, where mass tourism should be avoided.
- All parties concerned, including in particular the private sector, have a part to play in bringing about the sustainable development of tourism, and voluntary initiatives (codes of conduct, quality labels) should be encouraged.
- Particular importance should be attached to the local level, which is not only responsible for the sustainable development of tourism but should also derive particular benefit from tourism.

It will mark the successful beginning of internationally co-ordinated efforts to make tourism environmentally and socially sustainable so that many generations to come can continue to experience and enjoy the beauty of nature on our planet.

22

ECONOMICS AND ENVIRONMENT

Statistics change our view of the world. So statistics, however objective and accurate, are never value free but focus on what societies deem important. For better or worse, they guide government, business and individual decisions.

Until recently, the old game of India's economic growth was unquestioned and the score was kept between the national players by comparing their Gross National Product (GNP) or its narrower domestic version, Gross Domestic Product (GDP). It is time to take a closer look at the proliferation of new scoreboards, statistics and quality-of-life indexes which will redefine wealth and progress and change the future direction of human society.

Clarifying Values

These new scorecards and the 'greening' of GNP/GDP national accounts reflect the new 'green' accounting in thousands of balance sheets, reports and books on environments. At the very least, assumptions underlying old and new indicators are being clarified. The debate is still over what rather than how to measure, and what to do about values and amenities that are priceless.

The costs of GNP growth are now obvious—from felled forests, pollution exhausted soils, depleted natural resources and holes in the ozone layer to disrupt cultures and communities.

The concept of GNP/GDP was adapted into national accounting in India. With little re-examination, it continues to value bombs and bullets (defence expenditure), highly while setting the values of defence expenditure, education and public infrastructure—not to mention clean air and water and other environmental assets—at zero. It also ignores the some 50 per cent of production, which is unpaid—such as do-it-yourself home construction and repairs, food growing, household maintenance, parenting children and volunteering. In India such unpaid work can comprise up to 75 per cent of all production, particularly in agriculture sector.

Systems of National Accounts are based on GNP/GDP. Few economists, trade negotiators or development agencies questioned the basic assumption underlying it: that economies were generally in equilibrium, and that adding up a society's production and exchange of goods and services, measured in money terms, defined wealth and progress—however many social and environmental 'bads' came along with the 'goods'. Today's debates concern how best to calculate the costs of these 'bads' of production passed on to taxpayers or future generations. Some are easy to quantify: costs of cleaning up pollution can be calculated, and their increase marches in lock—step with the expansion of pollution control and environment industry sectors.

Confusing means with ends: Indian Government officials, business executives, academics and hundreds of thousands of civic organisations are beginning to agree that we have been confusing means (i.e. GNP growth) with ends (human development and the survival and further evolution of our species under drastically changed planetary conditions).

New environmental and resource realities, legislation and insurance liabilities are driving further overhauling of traditional accounts. There is a big issue over whether new indicators will be weighted in money terms to expand GDP, or whether the separate components—health, education, environment, etc. should be 'unbundled' so that the public can follow their own concern and hold politicians accountable for results:

Macroeconomists still try to expand GDP by pricing environmental amenities and costs. Social and natural scientists, while agreeing that environmental amenities must be valued at more than zero in GDP, advocate 'unbundled' physical indicators, such as water and air quality measures and rates of infant mortality. They suspect that economists 'contingent prices' for valuing the environment are theoretical and arbitrary.

Such 'Shadow prices' are derived by ecomomists from historic welfare theories and formulas based on 'willingness to pay' (WTP) of willingness to be compensated'. Thus, to arrive at a price for valuing a marshland (one of the most productive ecosystem on the planet), economists could poll voters and residents with no motives other than appreciation for marshes and their non-monetary or aesthetic values or their desire to preserve them and the rare species they might contain. Such contingent prices would be lower than those offered by a hotel developer with profit motives or by a biotechnology firm which had identified species in the area that could be used for pharmaceutical products. Worse, such pricing discounts poor people's needs and concerns, since they cannot afford to participate. Here the price system should be subordinated to more democratic decision-making, such as voting on whether or not to protect the marsh.

Economic Accountability

Most social and natural scientists, as well as voters, believe that economics must now take its place within interdisciplinary teams of statisticians from health, education, energy and environmental policy fields. Economics is not a science by rigorous standards, but a profession often lacking in the quality assurances and accountability that governs lawyers and doctors. GNP is a malfunctioning strand of our 'cultural DNA code' - carrying erroneous information and signalling to the body - politic a form of growth analogous to that of cancer cells which consume the host's body. The new national accounting methods being redesigned to correct or even replace GNP/GDP will function like healthy' cultural DNA strands', newly spliced in to govern healthier growth and more normal development patterns for human societies. Quantitative growth is dominant as children grow to adulthood, but once their mature size and weight are reached, this gives way to qualitative growth: education, social skills, broader awareness and even greater ethical understanding and wisdom. The statistical shift from GNP/GDP to sustainable development indicators mirrors such maturing of societies, recognising new goals and the traits human beings must now rapidly develop if we are to restructure our society for sustainability.

The new scorecards allow Indians to move beyond economism and ideologies of left and right to measure results directly and hold our business and government leaders accountable for implementing progress on the major goals of individual voters, consumers and investors. The new scorecards can help broaden trade pacts to include sustainable development criteria.

23

POPULATION GROWTH AND ENERGY

It has been scarcely 200 years, the dawn of the Industrial Revolution—since humans abandoned sole reliance on firewood, other biomass fuels, and direct sunlight to meet daily energy needs. In the past half-century, global demand for energy grew twice as fast as population, as industrial nations burned coal, oil, and natural gas to fuel their economies. Over the next half-century, world energy demands are projected to continue expanding beyond population growth, as developing countries try to catch up with industrial nations.

Developing countries will see tremendous growth in energy consumption in the next half-century, as growing populations and increasing affluence combine to drive their energy demands to dizzying levels. Based on projections from the U.S. Department of Energy and the Intergovernmental Panel on Climate Change, total energy consumption in the developing world will grow by 336 per cent, nearly three times faster than population, over the next 50 years, from 3,499 million tons of oil equivalent to 15,255 million tons. By 2030, energy consumption in the developing world will likely surpass usage in industrial nations.

Rising per capita consumption accounts for nearly two-thirds of the growth in energy demand in poorer nations, but different population trajectories can have dramatic effects on future demands. For example, assuming the same growth in per capita energy demand, moving to the low U.N. population projection will reduce total energy demands from developing countries by 2,792 million tons of oil equivalent the output of nearly 3,000 average-sized coal-fired power plants.

In the next 50 years, the greatest growth in energy demands will come where economic activity is projected to be highest: In Asia, where consumption is expected to grow 361 per cent, though population will grow by just 50 per cent. Energy consumption in Latin America and Africa is projected to increase by 340 per cent and 326 per cent, respectively. Lower rates of population growth in Asia, compared with Latin America and Africa, mean that energy use per person will increase most in Asia. Nonetheless, in all three regions, local pressures on energy sources, ranging from forests to fossil fuel reserves to waterways, will be significant.

When per capita energy consumption is high, even a low rate a population growth can have significant effect on total energy demand. In the United States, for example, where current per capita energy demand is nearly double that in other industrial nations and over 13 times that in developing countries, the 75 million people projected to be added in the next 50 years will boost energy demands by 758 million tons of oil equivalent, roughly the same as the present energy consumption of Africa and Latin America.

World energy use per person doubled between 1950 and 1973, before confronting a short-term slowdown when restricted exports from oil-producing nations drove up energy prices. Another price shock, combined with a global economic recession, resulted in the slowdown of the early 1980s. The most recent stumbling block in energy growth followed the 1989 revolution in Eastern Europe, when energy use in the former Soviet states plummeted. Although DOE and IPCC project substantial future growth, similar forces may act to check such a development.

World oil production per person reached a high in 1979 and has since declined 23 per cent. Moreover, estimates of when global oil production will peak range from 2011 by petroconsultants to 2025 by the IPCC, signaling future price shock as long as oil remains the world's dominant fuel.

Although people born in 1950 saw per capita oil production quickly double in a few short decades, those born in 2000 are likely to see it cut in half, dropping below 1950 levels.

In addition, meeting increased energy demands will require more storage an transportation infrastructure. Communities without a reliable supply of clean water or an adequate system for waste disposal may also fall short in connection to power supplies. For the estimated 2 billion who are still off the grid, and also experiencing high rates of population growth, decentralized energy technologies, such as solar roof shingles and fuel cell power generators, are likely the most feasible and affordable option for meeting increased energy demands.

Yet it will not necessarily be the scarcity of fuel that constrains future growth in energy consumption, but rather concerns about climate change, air quality, and water quality. Growing climate concerns will require massive reductions in fossil fuel use at a time when demand for energy is soaring. A shift to renewable energy sources, such as solar energy and wind power, in addition to continued efficiency gains for power plants, cars, and appliances, holds great promise for meeting future energy demands without adverse ecological consequences.

24

POPULATION GROWTH AND URBANISATION

The world's cities are growing far faster than its population. Indeed, aside from the growth of population itself, urbanisation is the dominant demographic trend of the half-century now ending. In 1950, 750 million of the world's people lived in cities. By 1996, this had at least tripled, to more than 2.6 billion. The number projected to live in cities by 2050, some 6.5 billion people, exceeds world population today.

Urbanisation on anything like the scale that we know today is historically quite recent. In 1800, only one city, London, had a million people. Today, 326 cities have at least that many people. And there are 14 mega cities, those with 10 million or more residents. Tokyo is the largest at 27 million. Mexico city is second, at 17 million. New York city and Sao Paulo are close behind, with 16 million each. Rounding out the list in descending size are Bombay (15 million), Shanghai (14), Los Angeles (12), Calcutta (12), Buenos Aires (12), Beijing (11), Osaka (11), Lagos (10), Rio de Janeiro (10), and Delhi (10).

The rate of growth of cities in industrial countries during the first century or so of the Industrial Revolution was relatively slow. Today's cities are growing much faster. It took London 130 years to get from 1 million to 8 million. Mexico city made this jumps in just 30 year.

Measured in annual growth, some cities, such as Lagos, Nigeria, are growing at 5 per cent a year; Bombay is growing at nearly 4 per cent. The world's urban population as a whole is growing by just over 1 million people each week. This urban growth is fed by natural increase of urban populations, by net migration from the countryside, and by villages, by net migration from the countryside, and by villages or towns expanding to the point where they become cities or they are absorbed by the spread of existing cities.

During the early stages of industrialisation, urbanisation was largely in response to the pull of employment opportunities in cities. More recently, however, the movement from countryside to city has been more the result of rural push than of urban pull. It is a reflection of the lack of opportunity in the countryside as already small plots of land are divided and then divided again with each passing generation, until they become so small that people can no longer make a living from them.

Historically, cities and the surrounding countryside had a symbolic relationship, with the latter supplying food and raw materials in exchange for manufactured products. Today, cities are tied much more to each other and to the global economy. The food and fuel that once came from the surrounding countryside now often comes from distant of the planet.

As societies urbanize, the use of basic resources, such an energy, and water, rises. In traditional rural societies, for example, people live on the land and thus do not need to travel to work. But once they migrate to cities, commuting becomes the rule, not the exception. In villages, most of the food that is consumed is produced locally, requiring little energy for processing, packaging, and transportation; once people move into cities, on the other hand, virtually all their food must be brought in. In a village where residents typically draw their water from a central well and carry it to their homes, water use in necessarily limited. But when villagers move to urban high-rise apartment buildings with indoor plumbing, replete with showers and flush toilets, water consumption soars.

The ecology of cities is a continuing challenge to city managers simply because cities require the concentration of huge quantities of water, food, energy and raw materials. The waste products must then be dispersed or the city will become uninhabitable. As cities become larger, the disposal of residential and industrial wastes becomes ever more challenging.

Partly as a result of the mounting pressure for people to migrate to cities, the growth in urban populations is far out-stripping the availability of basic services, such as water, sewerage, transportation, and electricity. As a result, life in urban shantytowns is plagued by poverty, pollution, congestion, homelessness, and unemployment.

Since the beginning of the Industrial Revolution, the terms of trade between countryside and city have favored the latter simply because cities control the scarce resources in development, namely capital and technology. But if the price of food rises in the years ahead, as now seems likely, the terms of trade could shift, favoring the countryside. If in the new world of the twenty-first century the scarce resources are land and water, those controlling them could have the upper hand in determining rural/urban terms of trade.

This aside, if recent trends continue, within the next several years more than half of us will be living in cities—making the world more urban than rural for the first time in history. We will have become an urban species, far removed from our hunter-gatherer origins.

25

POPULATION GROWTH AND WASTE

A growing population increases society's disposal headaches—the garbage, sewage, and industrial waste that must be gotten ride of. Even where population is largely stable-the case in many industrial countries—the flow of waste products into landfills and waterways generally continues to increase in coming decades, as they will in many developing countries, mountains of waste will likely pose difficult disposal challenges for municipal and national authorities.

Data for waste generation in the developing world are scarce, but citizens in many of these countries are estimated to produce roughly half a kilo of municipal waste each day. If this figure is applied to today's population, a total of 824 million tons of municipal waste is being churned out annually in developing countries. Population growth alone would boost this number to 1.5 billion tons by 2050. But waste rates tend to climb with rising incomes; a developing world generating as much waste per capital as industrial countries do today would be producing some 3.6 billion tons of municipal waste in 2050. Moreover, prosperity boosts the volume of waste as the share of plastics, metal, paper, and other non-organics rises.

Local and global environmental effects of waste disposal will likely worsen as 3.4 billion people are added to global population over the next half-century. Acids from organic wastes, for example, and poisons from hazardous wastes often leach from landfills, polluting local groundwater supplies. And rotting organic matter generates methane, a greenhouse gas.

If the waste is incinerated rather than thrown into landfills, cities will have to worry about increases in cancer-causing dioxin emissions, one of the byproducts of burning garbage.

Meanwhile, today's largely unmet sanitation needs could also be greatly exacerbated by population growth. Half the world's people do not have access to a decent toilet, according to UNESCO and the World Health Organisation. Lack of sanitation is a leading cause of disease: WHO reports that half the developing world suffers from one of the six diseases associated with poor water supply and sanitation. One of these, diarrhea, is the biggest killer of children today, taking an estimated 2.2 million young lives each year. Unless the expected growth in population of the developing world is matched by an increased commitment to provide adequate sanitation, these health problems are likely to expand.

While the greatest shortage of sanitation is found in rural areas, the need is most urgent in cities, because of the greater potential there for pathogen-tainted water to sicken people on a massive scale. This urban need poses a particular challenge, because the ranks of city dwellers will swell in the next century. In contrast to global population, which is projected to increase by 54 per cent over the next half-century, cities will see much greater growth—about 128 per cent. Developing-country cities which failed to meet the sanitation needs of more than a half-billion residents in 1994, will be hard-pressed to service the more than 3 billion people who will be added to cities in the next 50 years.

Prospects for providing access to sanitation are dismal in the near to medium term. Just to keep from losing ground, the rate of provision of service to urban dwellers needs to more than double in Asia. In Africa, it would have to increase by 33 times. And to achieve full coverage by 2020, service provision would have to triple in Asia and increase by 46 times in Africa. Despite the attention focused on sanitation, governments have not demonstrated the will to meet this growing challenge.

26

USING ECONOMICS TO ADVANTAGE

In the eyes of the public, the economic sectors, for instance energy, transport and agriculture, are often seen as pursuing interests that conflict with environment and health. They are the originators of pollution and often devise economic arguments to oppose changes in their practice that could improve environment and health. This behaviour has led the public, as well as environment and health professionals, to view economic analysis negatively. However, these economic arguments are often inadequate and unconvincing from the point of view of many economists.

In fact, the economic rationale is bound to reflect as closely as possible the preferences of the population and thus to take much greater account of environment and health. If used by environment and health authorities, economic analysis can be turned into a powerful tool for supporting their policies.

Why Use Economics?

First, economies can help to make explicit the benefits of environmental health improvements and the costs of the impacts. This provides additional arguments to encourage decision-makers to integrate environment and health considerations in their policies.

Second, current prices rarely reflect the full environment and health costs of the production or consumption of goods and services. Therefore, producers and customers have no economic reasons to reduce the impact they have on environment and health, as they do not pay prices that reflect this impact. Nor are they encouraged to take it into account in their investment decisions and lifestyle choices.

This could be corrected by reflecting as much as possible environment and health costs in the prices. Economic instruments, such as environmental taxes or tradable permits are a promising solution. A first step in that direction is the removal of subsidies that support practices harmful to the environment and health. In most of the cases, however, it would be difficult to remove decorative subsidies immediately and charge the full amount of environment and health costs. Nevertheless, negotiating plans and timetables to do so progressively, is a strong signal to the economic actors. It modifies their anticipation of future prices, as they know they will have to pay in the future for the environment and health costs they will create. This drives them increasingly to design their long-term choices and strategies in an environment-friendly way.

Finally, the setting of new economic instruments is usually under the responsibility of the Ministry of Finance. It also implies negotiations with economic sectors. Therefore environment and health authorities will need to play a more pro-active role in order to advance the integration of environmental health in sectoral and economic policies. Success will depend on their ability to discuss and present economic arguments in support of environmental health considerations.

A Promising Initiative

The present situation is that many environment and health authorities have few skills in using economic arguments and that economic sectors very often continue to ignore environment and health considerations.

International organisation—will also be invited to strengthen their co-operation in environment and health economics. In order to sustain the policy changes, promoting environment and health, co-operative efforts will aim.

- To support the development of the capacities of the environment and health authorities to use economic analysis;
- To improve the focus on health outcomes in national or inter-country processes dealing with environment and health issues. This will include the contribution of health expertise in these processes and the use of economic arguments to greater advantage;
- To exchange information early in the planning process of their respective programmes that use economic tools for addressing environment and health.
- To further co-ordinate their current and future activities in support of environment and health.

WATER:
Will be there Enough?

Humankind has a special relationship with water. In every civilization, the most ancient traditions associate this precious resource with the origins of life, purification and regeneration. Far from being a mere raw material such as oil, water is vital for life, indispensable to the economy and so rich in symbolic value that it triggers passionate responses. All the computers in the word will never be able to express the real perception of the value of water or codify the interactionss between it and peoples.

For decades, experts have been making grim forecast that the Earth will start running out of water and that conflicts over this precious liquid will erupt into wars. The situation is indeed alarming.

How much water is there in the earth's reserves? Highly expensive probes have been sent to the moon, mars and the satellites of Jupiter and Saturn to findout whether there is water on them, but we still lack accurate data about the earth's hydrological resources. Such information would help to provide a cleaner picture of the future and, especially to foresee the global repercussions of demographic and climate change.

One thing we know about water is that there is plenty of it. The total volume is put 1.4 billion cubic Kilometers—which could be imagined as a 2,650-metre-deep layer of liquid evenly disturbed over the entire surface of the planet. But 98 per cent of it is salt water, mainly in the oceans and seas.

Most of the earth's fresh water is trapped in the polar ice caps, less than 1 per cent of it is available in lakes, rivers and shallow, easily-accessible acquifers. These water resources are constantly in flux. Water from the oceans and land evaporates into the atmosphere before falling again as rain or snow, nourishing plants and swelling rivers that flow into the sea. It also seeps through the ground and percolates down to acquifers. Very deep groundwater, known as fossil water, is impervious to seepage and not renewable

In the industrialised countries, all you have to do is turn a tap and before you know where you are you've used a considerable amount of water up to 600 litres per person a day in the United States. In hot developing countries, where shanty-towns on the edges of cities are crowded with growing numbers of migrants from the rural areas, a spigot and two litres of water a day are a luxury.

Over 1.3 billion people received improved drinking water services and some 750 million got better sanitation facilities during the International Drinking Water Supply and Sanitation Decade (1980-1990). Approximately 1.2 billion people still have no access to drinking water and 2.9 billion lack sanitation. The resulting water-borne diseases take the lives of five million people a year, most of them children.

Farming and manufacturing account for most of the world's water consumption, far outdistancing human needs of the 3,240Km^3 of fresh water drawn every year, only 8 per cent are used for human consumption. Each year fewer than ten countries use 60 per cent of the world's 40,000 billion m^3 of surface and ground water. Lastly, per capita consumption rises with the standard of living, ranging from 260 litres a day per person in Israel to 200 in Europe, 70 for a Palestinian on the West Bank and 30 in Africa.

The Dangers of Irrigation

Demand runs highest in places where irrigation is indispensable, such as central Asia, Iraq, Iran, Pakistan Madagascar and also in some industrially developed countries such as the United States. Farming accounts for two-thirds of the total water resources used by humans—a figure that rises to 80 per cent in the Southern countries. Developing countries consume twice as much water per hectare to irrigate land as industrialised nations, yet their production is three times lower.

Because of the heat, half the water, evaporates in storage areas or when flowing through open-air irrigation canals. Poorly conceived irrigation projects lead to deterioration of the soil.

The first example is Pakistan, during the first half of the twentieth century, 10 million hectares were abundantly irrigated in the Indus plain. Water-logging caused by irrigation, combined with a high rate of evaporation, has led to salinisation of the soil, making it unproductive. The second example, the Aral Sea in the former Soviet Union, is different but the result is the same. Much of the water from the Syrdarya and Arnu Darya rivers that flow into the huge lake has been diverted to feed 1,80,000 kms of irrigation canals, only 12 per cetn of which have been made watertight. The rivers' flow is considerably restricted and the Aral Sea is drying up. Irrigation for agricultural purposes is expensive. To make it profitable farmers must use massive amounts of pesticides, herbicides and fertilizer on increasingly exhausted soil. The impact of pesticides on health has been overlooked. The child morbidity and mortality rates are among the world's highest.

28

SOLUTIONS FOR A WATER-SHORT WORLD

As populations grow and water use per person rises, demand for fresh water is soaring. Yet the supply of fresh water is finite and threatened by pollution. To avoid a crisis, many countries must conserve water, pollute less, manage supply and demand, and slow population growth.

Caught between growing demand for fresh water on one hand and limited and increasingly polluted water supplies on the other, many developing countries face difficult choices. Populations continue to grow rapidly. Yet there is no more water on earth now than there was 2,000 years ago, when the population was less than 3 per cent of its current size. Rising demands for water for irrigated agriculture, domestic (municipal) consumption, and industry are forcing stiff competition over the allocation of scarce water resources among both areas and types of use.

Today 31 countries, accounting for under 8 per cent of the world population, face chronic fresh water shortages. By the year 2025, however, 48 countries are expected to face shortages, affecting more than 2.8 billion people–35 per cent of the world's projected population. Among countries likely to run short of water in the next 25 years are Ethiopia, India, Kenya, Nigeria, and Peru. Parts of other large countries, such as China, already face chronic water problems.

In much of the world polluted water, improper waste disposal, and poor water management cause serious public health problems. Such water-related diseases as malaria, cholera, typhoid, and schistosomiasis harm or kill millions of people every year. Overuse and pollution of water supplies also are taking a heavy toll on the natural environment and pose increasing risks for many species of life.

What Can Be Done?

It may already be too late for some water-short countries with rapid population growth to avoid a crisis. Many other countries can avoid the coming crisis if appropriate policies and strategies are formulated and acted on soon. Whether water is used for agriculture, industry, or municipalities, there is much room for conservation and better management. Effective strategies must consider not only managing the water supply better but also managing demand better.

To avoid catastrophe over the term, it also is important to act now to slow the growth in demand for fresh water by slowing population growth. Currently, in many developing countries millions of people want to plan their families and to use contraception. Family planning programmes have played an important role in assuring individual reproductive health and in reducing national fertility levels. Continuing and expanding these programmes also can help assure that population growth eventually slows to sustainable levels in relation to the supply of fresh water.

Towards a Blue Revolution

The world needs a Blue Revolution to conserve and manage fresh water supplies in the face of growing demand from population growth, irrigated agriculture, industries, and cities—just as the Green Revolution transformed agriculture in the 1960s. A Blue Revolution will require coordinated responses to problems at local, national, and international levels.

Locally led initiatives show that water can be used much more efficiently. When communities manage fresh water resources efficiently, they also manage other natural resources better, improve sanitation, and reduce disease. At the national level, especially in water-short regions with dense populations, adopting a watershed or river-basin management perspective is a needed alternative to uncoordinated water-management policies by separate jurisdictions. At the international level countries that share river basins can fashion workable policies to manage water resources more equitably. Development agencies need to focus more on assuring the supply and management of fresh water resources and on providing sanitation as part of development and public health programmes.

A water-short world is an inherently unstable world. As the next century dawns, water crises in more and more countries will present obstacles to better living standards and better health and even bring risks or outright conflict over access to scarce fresh water supplies. Finding solutions should become a high priority now.

29

LIVING WITH LEVIATHAN

In the year 2015, there will mega-cities with more than 8 million inhabitants—22 of them in Asia. How will they cope? Humanity is about to set a new record. Nearly two-thirds of the planet's population will be living in cities by 2025, UN population experts say. Until now, rural people have outnumbered city-dwellers.

World population, according to the same projections, will top eight billion in 25 year's time, including five billion in cities. The increase will be particularly spectacular in the cities of the developing world, whose total population will double to four billion. We are going to see an unprecedented exodus of people from rural areas.

The demographer's predictions are only tentative of course. But the flow of people into megacities in developing countries is well under way. Several sociological changes are behind it.

Cities used to need muscle-power for the jobs they provided, the experts point out. But today they no longer attract people just because of their economic potential. There is plenty of evidence that they can go on steadily attracting people even when the job-generating sectors are in bad shape or disappearing.

People no longer move to urban centers because they are fairly sure to find work. They do so because they want to leave the countryside where there are too many people tilling the land and because they hope to leave poverty behind.

Rightly or wrongly, the city seems to offer progress and freedom, a vision of opportunity, an irresistible lure.

The result is that both inside and outside cities, there are more and more squatters and poor housing. Urbanisation in the developing world differs from that in the industrialized countries, in "the speed of the process, the growth of poverty, the extent of urban sprawl and the expansion of the informal economy."

How are the authorities in the developing world's urban areas responding to such "invasions"? In today's deregulated world, the trend is to question the very idea of providing the general population with basic urban services, most observers note. For want of resources, cities in developing countries are increasingly abandoning their public service function.

China is still an exception to this in several ways. Officials there, in a context of rigid planning—though this has eased in recent year—are trying to prevent the influx of more rural migrants than the city economies can cope with, as the example of Shanghai shows. Can such a policy, which works fairly well for the moment, survive the political and economic hangs under way?

At the other end of the scale is Lagos (Nigeria), whose expansion is chaotic, about 200 slums have sprung up in this African city. Every now and then, one of them is bulldozed without notice and without heed for its inhabitants. But Lagos survives, thanks to the vibrant ingenuity of its millions of citizens. Another revealing city is Jakarta, where the authorities themselves have joined in frantic property speculation. As a result of this speculation, more than 4.5 million people have been evicted from their homes in the last 30 years, with little compensation, to make possible the construction of high-rise blocks, which sometimes stand empty.

How do the original inhabitants of a city react to the massive influx of people from outside? In more and more cities, you see smart neighbourhood protected by guard—called "fortress-cities". In these fortified enclaves, built partly in response or real to imagined lack of security, the roads, sewage system, schools and other community services are private. Outside them, public areas have been abandoned to the least fortunate members of the society and the infrastructure there is crumbling or inadequate. The middle and poorer classes also defend themselves in their own neighbourhoods. One surprising case can be found in the satellite cities just outside Brasilia, where iron railing protect the houses, from fancy villas to the humblest shack.

Will the mega-cities of the 21st century be made up of island of "social tribes"- "anticities" of walled enclaves, whose wealthy residents refuse to pay taxes to provide facilities for the city's less fortunate inhabitants? Will cities still integrate their inhabitants?

"The existence of a slum means the authorities have failed." Says the World Bank. The bank encourage project where the state and the private sector join hands to help the less fortunate buy plots of land in areas with an infrastructure. Other experts say the "anti-social" aspects of globalisation should be blamed. They would like to see the big cities of the next century return to their original function as a crossroads and a meeting-place

30

AN AGENDA FOR CHANGE

The World's growing population, combined with unsustainable production and consumption patterns, is putting increasing stress on air, land, water, energy, and other essential resources.

- Development strategies will have to deal with the combination of population growth ecosystem health, technology, and access to resources. Meeting the unmet need for family planning and reproductive health services should be part of national sustainable development strategies.
- The world needs to do a better job of forecasting the possible outcome of current human activities, including population trends, per capita resource use, and wealth distribution.

Protecting the Atmosphere. The atmosphere is under increasing pressure from greenhouse gases that threaten to change the climate and from chemicals that reduce the ozone layer. Governments need to:

- Modernize existing power system to gain energy efficiency and develop new and renewable energy sources.
- Promote national energy efficiency and emission standards and develop efficient, cost-effective, and less polluting mass transit systems.

Combating Deforestation. Forests world-wide are threatened by uncontrolled degradation and conversion to other uses because of increasing human pressure.

- There is an urgent need to conserve and plant forests in developed and developing countries to maintain or restore the ecological balance and to provide for human needs.
- Governments need to work with business, scientists, local community groups, indigenous people, and the public to create long-term conservation and management policies for every forest region and watershed.

Sustainable Agriculture and Rural Development. Hunger is already a constant threat to over 800 million people, while the world's ability to continue meeting growing demand for food and other agricultural products over the long-term is uncertain. Soil erosion, salinisation, water-logging, and loss of soil fertility are increasing in all countries.

Agriculture has to meet rising needs mainly by increasing productivity, because most of the world's best croplands are already in use. At the same time further encroachment on land that is only marginally suitable for cultivation must be avoided.

- Sustainable agriculture and rural development will require major adjustments in agricultural, environmental, and economic policies in all countries and at the international level.

Conservation of Biological Diversity. The loss of the world's biological diversity continues, mainly from habitat destruction, over-harvesting, pollution, of foreign plants and animals (known and exotics). This decline in biodiversity is largely caused by human activity and represents a serious threat to our development.

- Develop national strategies to conserve and sustainably use biological diversity and to make these strategies part of overall national development efforts.
- Implement fair sharing of the benefits between providers and consumers of biological resources.

- Protect natural habitats. Promote the rehabilitation of damaged ecosystems.

Protecting and Managing the Oceans. Oceans are under increasing environmental stress from pollution over-fishing, and degradation of coastlines and coral reefs. About 70 per cent of marine pollution comes from sources on land. Countries should commit themselves to control and reduce degradation of the marine environment. They should:

- Build and maintain sewage—treatment systems and avoid discharging sewage near shell fisheries, water intakes and bathing areas.
- Develop land-use practices that reduce run-off of soil and wastes to rivers and thus to the seas. Use environmentally less harmful pesticides and fertilizers.
- Control and prevent coastal erosion and silting due to land uses such as unplanned construction.

Protecting and Managing Fresh Water. In many parts of the world there is widespread scarcity, gradual destruction, and increased pollution of freshwater resources. The causes include the inadequately treated sewage and industrial waste, loss of natural water catchment areas, deforestation and other chemicals into the water. The following approaches are key:

- The way to provide all people with potable water and basic sanitation is to adopt the approach "some for all rather than more for some". This approach can be achieved through low-cost services built and maintained at the community level.
- Nations need to identify and protect water resources and see that water is used on a sustainable basis. They need effective water pollution prevention and control programmes. There is a particular need for appropriate sanitation and waste disposal technologies for low-income, high-density cities.

31

SUSTAINABLE TOURISM AND THE ENVIRONMENT

Tourism is high on the international agenda. The 7th session of the Commission on Sustainable Development focused on tourism and subsequently work programmes on sustainable tourism are being developed. Also the Convention on Biological Diversity is embarking on tourism programmes and bilateral and multilateral financial institutions placed tourism high on their priority lists. The UN declared 2002 as the International Year of Ecotourism and the World Tourism Organisation adopted a Global Code of Ethics for Tourism at its General Assembly, held in Santiago de Chile.

The World Tourism Organisation forecasts that there will be 702 million international arrivals in the year 2002, that arrivals will top 1 billion in the year 2010 and that by 2020 international arrivals will reach 1.6 billion—nearly three times the number of international trips made in 1996, which was 592 million.

Travellers of the 21st century will go farther and farther. The Tourism 2020 Vision forecast predicts that by 2020 one out of every three trips will be a long-haul journey to another region of the world. It is expected that China will become a major force in international tourism and the WTO predicts that about 100 million Chinese will take international trips by 2020, thus putting them in fourth place in numbers of travellers after Germany, Japan and the United States. By the same time, China will attract 137 million visitors—63.5 million overseas visitors travelled to China in 1998 and thus outrank France as the world's top destination. It is estimated that during 1999 France will receive a record number of tourists of more than 70 million; in 2007 France hopes to attract 90 million visitors. The key resource for the most popular tourist destinations is the natural environment: coastal resort, tropical rainforests, wildlife in national parks and alpine skiresorts, all rely on a mixture of natural beauty, good weather and safe conditions to attract holiday destination is landscape and natural environment, followed by climate, the cost of the journey and the historical features of the place to visit. Hence, conserving the ecological integrity and environment is imperative if tourism is to be sustained.

The pressure from millions of tourists on water and marine resources, on land and landscape, on wildlife and habitat is enormous and often has devastating impact on the environment and the local population who are increasingly deprived of access to clean water and other natural resources.

In some regions, particularly in small island countries, tourism is one of the major reasons for wasting and polluting water: on average one tourist consumes at least 6 times more water than a local resident.

Major waters wasters and polluters are golf courses. In many countries, golf has brought heavy ecological and social costs: deforestation, the destruction of bio-diversity and erosion; dispossession of peoples' homes and farms; over-consumption and pollution of water and very high use of pesticides and fertilisers which threaten local residents, workers, wildlife and the golfers themselves. A survey by the Japanese National Doctors Health Insurance Association has revealed that many golfers, caddies and residents living near a golf course suffer from skin inflammation, disorders of the ear, nose and throat and other respiratory illnesses to the inhalation of pesticides because up to 90 per cent of the chemicals sprayed on golf courses end up in the air. In some areas in Thailand, diseases emerged which, prior to the construction of golf courses, had not been known.

In some regions, golf courses have depleted water supply, agricultural production has come to a halt, peasants have become impoverished and forced to migrate to urban areas in search of employment. Golf courses take large amounts of land. It is estimated that each year worldwide up to 5,000 hectares of forest are cut to clear land for golf courses.

Very often, the construction of golf courses forms an integral part of a comprehensive tourism project. Adjacent to the golf course, condominiums and/or hotels are built, very often also a marina, an airport and a casino. Studies have shown that such a complex not only has touristic objectives but is often connected to drug trafficking and money-laundering. Even the US State Department has emphasised the link between tourism, money-laundering offshore banking.

Cruise ships are a major cause for pollution in the Caribbean, destroying maritime life and reefs by releasing waste into the ocean. Recently the Royal Caribbean, the world's second largest cruise line was fined a record sum of US$18 million for dumping waste oil and hazardous chemicals into the sea. The company admitted to routinely dumping wasted oil from its fleet and that it deliberately dumped in U.S. harbours and coastal areas many other types of pollutants, including hazardous chemicals from photo processing equipment, dry cleaning shops and printing presses. Some hazardous materials, including toxic solvent from dry cleaning operations, were illegally placed in the garbage aboard the ships. The material was then either incinerated on the ship or dumped in U.S. or foreign ports mixed with ordinary garbage.

It was announced that the Royal Caribbean Cruise reported a profit of US$338 million in 1997, a 93 per cent increase over the previous year, Carnival Corporation's Holland, the biggest cruise company with a turnover of US$ 3 billion in 1997 made a net profit of US $836 million, 25 per cent more than in 1996. Both cruise companies have recently been fined millions of dollars for dumping untreated bilge water, oil and other waste into Alaskan waters.

However, the impact of oil and hazardous waste on water, maritime life and coral reefs is devastating and all fines paid for the damage caused by the cruise ships will not revive dead corals.

A recent Green peace study on coral reefs—one of the marine world's great natural treasures—predicts that the coral bleaching which dramatically whitened many of the world's reefs last year will escalate rapidly under accepted global climate models and that the damage would wreak havoc in fisheries and tourism, disrupting the economies of many nations.

A WWF study recently published on "Climate Change and its Impacts on Tourism", warned that droughts, rising seas, flash floods, forest fires and diseases could turn profitable destinations into holiday horror stories. The report urges the tourist industry to persuade western industrialised governments to take more concerted action to reduce their nations' carbon dioxide emissions the main cause of global warming.

The Need for Action and Education

If governments, the international community and the tourism industry want to save the world's major tourist destinations, immediate action is required. Governments and the tourism industry must abide to the principle that environmental protection is an integral part of tourism development. In order to protect the environment and mitigate the damages caused by tourism, some countries have decided to take action: The Spanish Island Minorca and the Seychelles will introduce Eco-tax on tourism. This tax will be around US$12 per person in Minorca and its revenues are earmarked for the maintenance of national parks and the restoration of damaged coastline. Visitors to the Seychelles will have to buy a so-called "gold-card" at a price of $100 which entitles unlimited access to the country; income from this card will be used for sewage management and protection of fresh water supply.

Only if tourism investors and developers:

(a) consider the natural capacity for the regeneration and future productivity of natural resources.

(b) recognise the contribution that people and communities, customs and lifestyles, make to the

tourism experience and therefore accept that these people must have an equitable share in the economic benefits of tourism; and

(c) listen to local people in the tourist destinations, tourism may become sustainable.

Education and awareness raising campaigns at all levels are therefore imperative.

32

BIG-DAM CONSTRUCTION IS ON THE RISE

As the era of big dams fades in North America, construction is increasing in Asia, fueled by growing demand for electricity and irrigation. Worldwide, the number of dams under construction in 1998 (the latest year for which global data are available) rose by more than 9 per cent, to at least 1,240; after a much smaller increase in 1992. The Table 32.1 shows that China accounts for more than one fourth of the big dams under construction, while China, Japan, South Korea and India together account for more than half. The increases follow a decline in the 1980s, when construction worldwide averaged less than half that of the preceding 25 years.

Data for the early 1990s, though incomplete, indicate a shift toward larger dams, with concomitant greater sccial and economic costs. In 1992, 60 per cent of the dams being built were more than 30 meters high compared with only 21 per cent of existing dams in 1986. Construction of dams higher than 100 meters rose by some 27 per cent between 1991 and 1993, half of these large structures were built by just three countries—Japan, China, and Turkey.

Dams are a symbol of modernity and a source of national prestige, partly because they are a multipurpose tool of development. They generated more than 18 per cent of the world's electricity in 1998, and reservoir water irrigates millions of hectares of land around the world, raising agricultural yields two to three times over those of dry land farming, and in many cases bringing agriculture to regions that could not otherwise support it. Areas with a high percentage of irrigated agriculture, such as China, India, and the western United States, rely heavily on reservoirs for water. Irrigation is increasingly listed as the primary purpose of new dams, possibly reflecting growing food needs and water scarcity in many developing countries.

Dam reservoirs also protect societies against drought. Water in the world's reservoirs effectively increases the normal supply from rivers by some 30 per cent. This insurance can be invaluable; Egyptians maintain that the Aswan Dam saved Egypt from catastrophe during the drought of 1979-80.

In the past two decades, however, as it became clear that dams have serious costs as well as benefits, dam construction began to decline. Loss of land to reservoirs can be significant. The Narmada Sagar Project in India is expected to drive 31 species of plant to extinction because of habitat changes caused by reservoir creation. Decomposing plant life at reservoir bottoms can release as much methane and carbon dioxide both greenhouse gasses—as a coalfired plant with the same electricity-generating capacity, according to the Freshwater Institute in Canada.

Dams also restrict stream-flow with often disastrous biological and economic consequences. Populations and fisheries have been nearly eliminated in many places worldwide, as on the Columbia River in the United States, where once plentiful salmon are now endangered species, and valuable fisheries have collapsed. Charges in temperature and water alter habitats of fish and other wildlife. Sediment that would normally enrich flood plain soils gets trapped behind dams. This last problem, siltation, can shorten a dam's useful life; in an extreme case, the turbines of China's Sanmenxia Dam were shut down in 1964 after only four years of operation when the dam's reservoir filled with sediment.

By restricting stream-flows, dams can also contribute to the decline of local economies. One example is Indian diversion of the Ganges river before it reaches Bangladesh, which has caused $25 million in losses for an agricultural project in northwest Bangladesh, and increased salination of fishing areas at the Bay of Bengal, the river's natural outlet.

Dams exact a high human toll as well. Millions of people were resettled in the past half century to make room for dam reservoirs. China's three Gorges Dam alone may well displace more than one million people. Those affected have had little say in their resettlement, and most displaced citizens wind up in worse situations than before their move.

Dams can increase health risks. Waterborne disease, including malaria and river blindness, have been introduced and spread as reservoirs are created. Large-scale water

Table 32.1: Leading Builders of Big Dams
(Higher than 10 meters)[1]

Country	New Starts[2] (Number)	Under Construction (Number)
China	82	311
Turkey	84	190
Japan	11	140
South Korea	2	125
India	48	76
United States	30	55
Spain	16	53
Romania	0	39
Italy	0	37
Tunisia	16	28
Algeria	6	27
Iran	1	76
Thailand	7	17
Greece	3	14
France	8	12
Brazil	4	12

projects were a major contributor to the 75 per cent global increase in cases of schistosomiasis, sometimes fatal disease.

Advocates of dams acknowledge these shortcomings, but argue that these effects can be mitigated, or that the alternatives to dams would be worse. Others argue that evaluation of dams involves widely differing environmental, social, and political contexts, and a project-by-project analysis of each dam is needed.

The World Bank—strong supporter of big dams—appears to have slowed its involvement. The Bank was involved with an average of 18 dam projects a year between 1980 and 1985, but with only 6 a year between 1986 and 1993. In fact, the Bank established a commission in 1994 to hear complaints from parties affected by proposed dam projects, the first of whom were people affected by the Arun III dam in Nepal.

The future of dams is unclear. Proponents note that only a fraction of the technically usable hydropower potential in developing countries has been harnessed. But the discussion has moved beyond technical considerations. Wide ranging development, conservation, energy use and population considerations are or should be part of every proposed dam's calculus. It remains to be seen whether current and future big dams can meet the increasingly stringent environmental and social standards expected of them, and whether development needs can be met in other ways.

33

POPULATION GROWTH AND FRESH WATER

Wherever population is growing, the supply of fresh water per person is declining. As a result of population growth, the amount of water available per person from the hydrological cycle will fall by 74 per cent between 1950 and 2050. Stated otherwise, there will be only one-fourth as much fresh water per person in 2050 as there was in 1950. With water availability per person projected to decline dramatically in many countries already facing shortages, the full social effects of future water scarcity are difficult even to imagine. Indeed, spreading water scarcity may be the most underrated resource issue in the world today.

Evidence of water stress can be seen as rivers are drained dry and as water tables fall. The Colorado River in the south-western United States now rarely reaches the sea. The Yellow River, the northernmost of China's two major rivers, has run dry for a part of each year since 1985, with the dry period becoming progressively longer. In 1997, it failed to make it to the sea for 226 days. The Nile, the largest river in the Middle East, has little water left when it reaches the sea.

Water tables are now falling on every continent, including in major food-producing regions. Among those where aquifers are being depleted are the U.S. southern Great Plains the North China Plain, which produces nearly 40 per cent of China's grain; and most of the India. Wherever water tables are falling today, there will be water supply cutbacks tomorrow, as aquifers are eventually depleted.

Some 70 per cent of the water pumped from underground or diverted from rivers is used for irrigation, 20 per cent is used for industrial purposes, and 10 per cent is for residential use. Water use patterns vary widely by region. In Europe, for example, where agriculture is largely rainfed, water withdrawals are dominated by industrial use. In Asia, in contrast, irrigation accounts for 85 per cent of all water use.

As countries press against the limits of their water supplies, the competition among sectors intensifies. The economics of water use does not favor agriculture. One thousand tons of water can be used to produce one ton of wheat worth $200 of to expand industrial output by $14,000. This ratio of 70 to 1 explains why industry almost always wins in the competition with agriculture for water.

As the growing demand of water collides with the limits of supply, countries typically satisfy rising urban and residential demands by diverting water from irrigation. They then import grain to offset the loss of irrigation water. Since it takes at least 1,000 tons of water to produce a ton of grain, importing grain becomes the most efficient way to import water. North Africa and the Middle East—a region where population growth is rapid and every country faces water shortages—has become the world's fastest growing grain import market during the 1990s. In 1997, the water required to produce the grain and other foodstuffs imported into the region was roughly equal to the annual flow of the Nile River.

In both China and India, the two countries that together dominate world irrigated agriculture, substantial cutbacks in irrigation water supplies lie ahead. The combination of the effects of aquifer depletion in key countries such as these and the growing diversion of irrigation water to nonfarm uses in many countries makes it unlikely that there will be much, if any, increase in total irrigated area over the long-term. Already the irrigated area per person has been slowly declining since 1978, falling from a historical high of 0.047 hectares per person to 0.045 hectares in 1996—a drop of 4 per cent. If the total irrigated area remains at roughly 263 million hectares until 2050, this key figure will fall to 0.028 hectares per person in 2050—declining by an additional 38 per cent. Such a shrinkage will pose a formidable challenge to the world's farmers.

About a billion people will be living in countries facing absolute water scarcity by 2025. These nations do not have enough water to maintain 1990 levels of food production per person from irrigated area, even with high irrigation efficiency, and to meet the needs for domestic, industrial, and environmental purposes as well. They will have to reduce water use in agriculture in order to satisfy residential and industrial water needs. The resulting decline in domestic food production will force them to import more food, assuming it is available. Although detailed water projections by sector for each country are not available for 2050, the number of water-deprived people will be far greater than in 2025 if the world continues on the U.N. medium population trajectory. The bottom line is that if we are facing a future of water scarcity, then we are also facing future of food scarcity.

34

POPULATION GROWTH AND NATURAL RECREATION AREAS

Population growth during the past 50 years has made it difficult to set aside and conserve natural areas. Another half century of growth will put even more pressure on protected areas as formerly small, distant settlements encroach on these sites and as the number of people (both local and visitors) who use these sites explodes.

National parks, forests, wildlife preserves, beaches, and other protected areas offer sanctuary to various habitats and indigenous communities, in addition to providing resources for local peoples. In an urbanizing world, these sites provide an opportunity for healthy interaction with the natural environment, as well as rare serenity.

From Buenos Aires to Bangkok, dramatic population growth in the world's major cities and the sprawl and pollution they bring threatens natural recreation areas that lay beyond city limits. Tremendous growth in the population of Bombay has already engulfed Borivili National Park, a reserve that was boyond the city's periphery only a decade age. With projected growth of 60 per cent in the next 20 years, Bombay may soon swallow up more distant areas. On every continent, human encroachment has reduced both the size and the quality of national recreation areas.

In nations where rapid population growth has outstripped the carrying capacity of local resources, protected areas become especially vulnerable. Although in industrial nations these areas are synonymous with camping, hiking, and picnics in the country, in Asia, Africa and Latin America most national parks, forests, and preserves are inhabited or used for natural resources by local populations.

An assessment by the World Conservation Union-IUCN of 30 protected sites in the developing world shows that these areas now act as magnets, attracting people to the rich oasis of water, fuel, food, and other resources they contain. Population growth rates in and around these areas are typically 2 percentage points above the national average—largely as a result of immigration from resource-starved areas.

As people seek out scarce resources, the resulting concentrations can be devastating. For example, population densities in the region surrounding Bwindi Impenetrable National Park in southern Uganda are some of the highest in all of Africa-exceeding 250 people for square kilometer. Though population at this site is expected to multiply, chronic land hunger already precipitates conflicts over fuel-wood collection, farming, cattle grazing, and bush burning.

Migration-driven population growth also endangers natural recreation areas in many industrial nations. Everglades National Park faces collapse as millions of new-comers move into South Florida.

Coastal recreation areas, including beaches, may be most burdened by the formidable combination of population growth and migration. All but one of the world's 15 largest cities, Mexico city, are coastal, and all of these cities will grow in the decades ahead. Whether it takes the form of expanding shantytowns in Kingston, Jamaica, or sprawling tract housing in southern California, virtually all the growth and movement in population in the next 50 years will occur in densely populated coastal corridors.

In nations already struggling to meet basic human needs, the prospect of establishing additional protected areas becomes increasingly slim. Throughout India, for example, while the national government designates areas as protected, state and local governments work to de-reserve these sites so that the resources can be harnessed to meet the needs of an additional 18 million Indians each year.

Sunbathers on beaches in Japan are often compared to sardines. People who use Central Park in New York city, which has nearly doubled in population since 1950, are faced with growing congestion and restrictions on activities. National Parks throughout North America are confronted with huge backlogs of requests to visit, having to turn tourists away. Tourism at Yosemite has boomed from roughly 4,000 visitors in 1886 to more than 4 million people (and their cars) today. It is often remarked that "Americans love their national parks to death", as increased visitation degrades campsites, trails, and wilderness.

Longer waiting lists and higher user fees for fewer secluded spots are likely the tip of the iceberg, as population growth threatens to eliminate the diversity of habitats and cultures, in addition to the space and quiet, that protected areas currently house.

35

DEVELOPMENT OF SERICULTURE

In majority of the developing nations, development efforts in the last decade were put on hold. In all the developing nations, poverty is on rise, economic growth has slowed down, employment has faltered and inflation is on upward swing. A greater proportion of the population in these countries depends on agriculture sector for their livelihood. But, in agriculture sector productivity is low. This is not only because of excess pressure on land but also agriculture in these developing nations is characterised by primitive technologies, poor organisation and limited capital.

Since the urban growth is severely limited, the growing labour forces have to get their employment in the rural areas or semi-urban areas in the coming decades. So, it has become compulsory for the agriculture sector to share the increasing burden since it is the major activity for the majority of the labour force.

It is obvious that in India larger than the necessary number of workers needed in agriculture are working on limited resources. As a result, agriculture sector has become an unprofitable activity with low levels of productivity. Moreover, the traditional crops have failed to absorb the growing labour force and to raise the incomes of the farmers above the subsistence level.

Since agriculture has been considered as the backbone of Indian economy, Indian agriculture can be broadly classified under two categories—(1) rainfed or dryland farming and (2) irrigated farming. Rainfed agriculture is primarily rain dependent. Rainfed agriculture in India supports 40 per cent of the total cultivated area. As irrigation facilities are inadequate, agriculture is still a gamble in the hands of monsoons.

For most of the irrigation projects, rainfall is the only source. Due to erratic nature of rainfall, most of the dams, tanks and other water reservoirs remain dry during summer season. These rainfed areas are frequently affected by periodic droughts, soil erosion, crop fluctuations and other related problems. To overcome these problems, the Government of India has implemented such projects which are helpful for eradication of poverty and unemployment and upliftment of the weaker sections.

In the context suitable strategies are needed to overcome the above mentioned problems. The discovery of productive employment opportunities in the integrated rural development assumes vital importance in the economic development of India. So, it is natural to start with agriculture as the biggest of India's industries. The most important method of securing an increase in agriculture output is by inducing the cultivators to adopt better agricultural practices.

In fact, Indian agriculture now is no longer confined to the cultivation of traditional crops. The farmers are encouraged to take up agriculture practices which are integrated with livestock culture, animal husbandry, dairying, fisheries, poultry, horticulture and sericulture to generate more income for each household.

In spite of the limitations in agriculture, it can be said without any doubt that Indian agriculture is on the threshold of entering into a stage of development characterised by a shift from static technology to a modern technology, in which capital requirement and purchased inputs occupy a larger share. But, much of the success of new programmes will depend upon the ability of the workers who act as growth promoters.

It is in this context, sericulture with its vast potential for employment generation in rural areas plays a vital role in alleviating rural poverty. However, this is another crop enterprise which is identified as one of the most appropriate labour-intensive cottage industries. This activity combines both agriculture and industry. It provides gainful employment not only at the stage of raising of mulberry plants but also at the stage of rearing of silk worms using output of the farmers as an input of the latter. Sericulture, for example, played a very important role in transforming the traditional bound Japanese agriculture into a modernised agriculture by intensive use of land, labour and capital.

Now, in India sericulture has become the most promising rural activity due to certain specific reasons like minimum gestation period, less investment, maximum employment potential and quick turnover of the investment. Sericulture generates direct and indirect employment in various ways. First, mulberry cultivation creates employment on the farm and secondly cocoon production which uses mulberry leaves as an input creates large-scale employment for the family labour of the mulberry growers, if that operation is also undertaken by the same household to reduce their underemployment in agriculture. Further, the reeling activity is also mainly undertaken in rural areas or semi-urban areas and the employment generated there would help to reduce the rural unemployment in a significant way. In short, sericulture as a whole, by its very nature of activities creates large scale employment and income generation opportunities in the rural and nearby semi-urban areas accelerating the economic growth of these areas.

Details of Sericulture Development in Kurnool District of A.P.

Sericulture is better suited for drought-prone areas. Kurnool district is one among the frequently drought affected district of Ryalaseema and suitably sericulture finds its place. Sericulture was first introduced in the district during 1975-76 in Bapananthapuram village of Atmakur Mandal. The area under mulberry was gradually increased over the years to 10,000 acres covering almost in all mandals in the district upto 1990 and the area stood at all-time high in 1990. But in the subsequent years the net area got reduced because of large scale uprooting by existing farmers due to losses suffered by them an account of steep fall in cocoon price and outbreak of deadly chronic disease in 1991-92 scanty erratic rains and continuous dry spell also major factors for uprooting. However, due to its lucrative income the area under mulberry is developed. Now the district is self-sufficient and even supplying seed cocoon to neighbouring districts. As for non-farm sector in sericulture in the district one 6 basin multiend silk reeling unit. One 6 basin silk reeling unit and 40 country charkas and a complex of 100 Twin Charkas are established in private sector. As many as 3,00 silk looms are functioning in Adoni area.

Infrastructure Facilities Available in the District

Govt. Sector

1.	Tech. Service Centers (Extension)	:	7 Nos. (Atmakur, Pamulpadu, M. Lingapuram, Nandyal, Nandikotkur, Adoni and Pathikonda.
2.	Tech. Service Centre (Non-Farm)	:	1 No. (Atmakur)
3.	Govt. Silk Reeling Unit	:	1 No. (Kurnool)
4.	Silk worm egg production centers	:	2 Nos. (Atmakur, Nandyal)
5.	Govt. Seed Farms	:	5 Nos. (Peapully, Kalichetla, Venkatapuram, Gajulapalli and Thangadancha)
6.	Seed Areas	:	2 Nos. (Adoni, Peapully)
7.	Govt. Seed Concoon Market	:	1 No. (Adoni)
8.	Govt. CB Cocoon Market	:	1 No. (Atmakur)

Private Sector

1.	Pvt. Charkas	:	4 Nos. Nandikotkur, Atmakur
2.	Silk Twisting Units	:	4 Nos. Yemmiganur, Kodumur Nandikotkur and Nandyal

Table 35.1

Item	*1994-95*	*1995-96*	*1996-97*	*1997-98*	*1998-99*	*1999-2000*	*2000 to 2001 as on 1/2001*
Area under mulberry cultivation (Acres)	4930	5273	5141	4232	5115	4195	3952
Cocoon production (in tonnes)	882	895	967	924	107	7.66	6.15
Qty. of raw silk produced booth in Pvt. and Govt. sectors (in Ton	2.23	2.2	3.8	6.0	3.3	3.3	4.3
	2450	2649	2774	2386	2558	2131	1935

1. Silk Power loom units : 1 No. Yemmiganur
2. Silk Looms: 3118 Nos. Kodumur, Yemmiganur, Adoni, Pathikonda, Koilakuntla Nandavaram
3. Silk Weavers Co-Op Society : 1 No. Gudikal (with 40 looms)

Details of Loans and Subsidies (Credit Flow)

Sl. No.	*Item*	*1996-97*		*1997-98*		*1998-99*		*1999-2000*	
1.	Loan	23.7	17.80	2.61	—	0.69	—	—	—
2.	Subsidy	8.78	7.13	5.12	—	0.46	—	—	—
3.	Margin Money	—	1.78	—	—	—	—	—	—
	Total	32.50	26.7	7.73	—	1.15	—	—	—

Achievements Under 8th Plan

Provided 100 House-cum-Work Sheds for Charka Silk Reeling to the rural poor people.

Proposals Under 9th Plan

(a) Providing rearing sheds on 50 per cent subsidy to Seed Farms by CSB

(b) Providing Drip Irrigation to Mulberry Gardens of sericulturists

(c) Establishment of Multiend Silk Reeling Machine's

Women Development Programme

Sl. No.	*Year*	*Programme*	*Physical*	*Financial*
1.	1995-96	1. Women Farmers Meet	350	
		2. Women Groups	02	
		3. Thrift-Cum-Credit Group	02	
2.	1996-97	1. Study Tours	125	0.4375
		2. Try in Mulberry Cultivation's & SW Rearing	25	0.1125
		3. Women Groups	02	0.50

3.	1997-98	1.	Try to women Sericulturists in New Technologies	75 —	0.225 —
		2.	Women Groups	01	0.25
4.	1998-99	1.	Trg. to Women sericulturists in New Techniques	30	0.30
		2	Women Groups	01	0.25
5.	1999-2000		Nil	—	—
6.	2000-2001		Nil	—	—

Trends in Employment Under Sericulture

Year	*No. of Persons engaged in sericulture*
1995-96	19076
1996-97	20564
1997-98	16928
1998-99	20460

Cocoons Markets in the District

1. Atmakur: : CB Cocoon Market
2. Kurnool : Notified Cocoon Market
3. Adoni : Seed Cocoon Market

Transactions Made in Govt. Cocoon Markets

Year	*Qty. of Cocoons sold*	*Market fee collected*	*Total value realised*
1995-96	5429	11904	566744
1996-97	6973	17518	822714
1997-98	11748	25892	1294000
1998-99	4458	8276	489000
1999-2000	137390	28999	1449886
2000-2001	1034	20463	1271947

36

THE DO'S AND DON'TS OF RISK REDUCTION

The most effective disaster mitigation measure that can be taken at community level is of people not to build in high-risk areas such as unstable slopes, riverbeds or flood plains.

Local knowledge about these hazards is usually good, especially among older people. Sometimes though, a hazard such as a geological fault is not obvious or visible, and surveyors and geologists have to be called in.

People can also ensure they reduce risks in their houses. Roofs should be secured against hurricanes. Roof shape is important for wind resistance. A flat roof is much more likely to be blown off than a pyramid-shaped one. Some worry about the cost of such measures, but they are no more than a small percentage of the total cost of the building and are well worth the investment.

Earthquake mitigation focuses on building codes, including correct use of steel and the strength of concrete mixes. Wooden buildings can be reinforced by braces and tying corners so as to make the structure react as a box.

Hazards in the home are not all structural. If you live in an earthquake-prone area, you should check the following:

- Are heavy objects like cabinets, TV stands and entertainment centres attached to the wall?
- Are heavy objects on the lowest shelves?
- Are water heaters and gas cylinders bolted to the wall?

- Do household members know how to turn off the gas, electricity and water supply?
- Are hanging objects such as fans and ceiling lights securely fastened?
- Are shelves fitted with wire or board to prevent objects falling off them during tremors?
- Are dangerous substances like fuel, poisons and chemical secured against spillage?
- Are plate-glass windows and doors covered with safety film to prevent shattering?

At the national level, governments should include mitigation in their disaster management policies. Zoning and land-use laws should ensure there are no buildings in an area likely to be flooded, say, once every thirty years. Golf courses and parks could be built there instead. Steep slopes would be left as wooded areas.

But planners and policy-makers do not usually have this freedom. Many rivers already flow through towns, so mitigation will take the form of reducing loss of life and damage after a flood. Ground floors can be designated non-sleeping areas, levees can be built and flood warning systems and evacuation plans can be developed.

Building codes should be drafted and where they already exist, should be reviewed and strengthened. They should ensure that buildings can survive a 7 magnitude earthquake or 200 km/h winds.

Standards of Safety

Many buildings were put up before codes were drafted, so they need to be "retrofitted" by strengthening. This is more expensive than building to resistant standards. Large public buildings and bridges are prime candidates for retrofitting. But small traditional buildings should be strengthened too. Much work has been done on this in India. Chicken mesh and mortar has been effective in reinforcing walls of adobe-type buildings against earthquake damage.

Emergency facilities such as hospitals, police stations and shelters, along with water, electricity and sewage systems, must also be able to remain functional after a disaster. They should be designed to a higher standard of safety than other buildings and retrofitted where necessary. This especially applies to hospitals—crucial after a disaster—as they can be knocked out of actions without structural damage. So non-structural mitigation measures are vital.

Cost is one of the reasons cited for lack of mitigation measures in poor countries. But one could say that such countries cannot afford not to take measures. Even reducing direct damage by one per cent through mitigation would have been worthwhile. But political support for mitigation is hard to drum up because little work has been done on quantifying the benefits of mitigation as opposed to cost.

Mitigation is also very important in the natural environment. Action taken in watersheds in the mountains will eventually affect the marine environment, especially in small countries where the distance between watershed and sea is small. Eroded soil washing down into the ocean results in silting which kills off coral and fish. Coral reefs help control beach and coastal erosion, and if they are damaged the coast is more vulnerable to flooding which will in turn cause more erosion. Coral reefs are also important fish hatching grounds, so loss of reefs will harm the fishing industry, especially in island states. Destruction of reefs and beaches also harms countries, which live off tourism. Chemical spills into waterways eventually reach the sea and also damage coral reefs and marine life. Damaged or stressed reefs are more vulnerable to natural hazards because they are less resistant to wave action.

Dialogue between environmental and disaster managers is essential. For mitigation to work, the entire society must be involved. Programmes should ensure local communities take part in planning at the same time as they ensure political and financial support at national and international level. Mitigation must be part of our daily life. If the present century "invented" mitigation and moved disaster management beyond mere response, the next must see that mitigation becomes an integral part of planning at all levels of society.

37

CONSUMING THE FUTURE

Now that we are to reach six billion of us, it is a good point to check again on what sort of lifestyles we pursue and what is the environmental impact of those lifestyles. It is curious that we have spent several decades being concerned about the growing numbers of humankind while not giving at least an equal amount of attention to the levels of living we aspire to, and how many natural resources we chew up thereby and how much population and waste we cause.

Everybody is a consumer of sorts. True, every fifth person scarcely qualifies for that designation, consuming goods worth less than $1 per day. Conversely, every seventh person qualifies for a designation of super-consumer, with a cash income at least fifty times greater. These latter are the people who, through their carbon dioxide emissions, are disrupting everybody's climate dozens of times more than the average citizen of the One Earth. Fair Play, anyone?

Much as the have-nots seek to match the have's, it is plain their efforts will not work out for a long time to come, at best. If every Chinese person were to consume just one additional chicken per year and if the said chicken were to be raised primarily on grain, this would account for as much grain per year as all the grain exports of the number two exporter, Canada. If the Chinese were to raise their per-capita consumption of beef, now only 4 kgs per year, to that of Americans, 45 kg, and if the additional beef were produced largely in feedlots after the manner of the United States, it would account for as much extra grain as the entire US grain harvest, less than one-third of which is exported. Because of its recent climbing up the food chain toward a meat-based diet, China has become one of the world's leading importers of grain. The global grain market today is around 200 million tons per year, and shows scant scope for significant increase.

As a further measure of its ambitions, the Chinese government has designated the auto industry as one of five industry "pillars". Today China has fewer cars than Los Angeles. If per-capita car ownership, together with oil consumption, were to match that of the United States, China would need 80 million barrels of oil per day—by contrast with the world's 1996 oil output of 64 million barrels of oil per day. The surge in carbon dioxide emissions would be unprecedented.

All this notwithstanding, there are already some 250 million newly affluent people in China. They are people with a household income equivalent to perhaps US$20,000, and enough discretionary income to enjoy the perquisites of the good life as perceived by these nouveaux riches. Top of the shopping lists are meat and more meat, followed by cars whether big or small. These are the badges of success: they show you have arrived.

The new consumers in China are matched by at least 200 million in India, and tens of millions in South Korea, Taiwan, Malaysia and Thailand (the recent economic setbacks have not permanently punctured the economic bubbles). Then there are 200 million more in Brazil, Argentina, Venezuela and Mexico, and more again in Hungary and other countries of Eastern Europe, also Turkey. Put them all together and they total about as many as the 800 million long established consumers in the ultra rich countries (the OECD grouping). When the current economic hiccups in Asia are left behind, the ranks of the new consumers can be expected to rise rapidly.

But they cannot hope to become super consumers. Where would all the extra gain come from? How could the global climate tolerate the huge additional pulse of carbon dioxide? There are all kinds of other environmental reasons to suppose that environmental constraints will become all the more constraining. True, technology could help moderate the environmental impact. We could enjoy twice as much material prosperity while using only half as much natural resources and causing half as much pollution and waste. But the new consumers will want to pursue the American dream to the hilt, and it is hard to see that the best technologies could enable huge numbers of affluent aspirants, perhaps two billion people by 2010, enjoying even half the material prosperity of Americans with average household incomes of $40,000.

But is it true "prosperity"—mental and emotional as well as material? Or is the American dream becoming a nightmare with its harried lifestyles and declining leisure time, where the shopping mall is the ultimate mecca, and the good life is a case of piling up goodies?

In any case, we cannot expect the new consumers to forego their "rightful share" of affluence unless the long-time affluent agree to cut back on their environmental ruinous lifestyles. It is these communities that must offer a strong example, and soonest. Where is the political leader who will espouse the new vision, however much it may be perceived as the ultimate vote loser?

38

CRISIS PREVENTION:
Can Better Development Planning Lessen the Toll of Civil Emergencies and Natural Disasters?

Even a cursory scan of the world's headlines is depressing: armed conflicts are grinding on in Somalia, Afghanistan and in a growing number of other countries. And the effects of natural disasters are becoming more catastrophic each year. International relief aid, in response to such emergencies, has increased substantially. But how large can these sums of money realistically be expected to grow? With no end in sight to the need for relief, the good-will of international donors is quickly giving way to disillusionment.

This leads us to a second question, which is, where does development fit in this grim scenario? For the development community to remain aloof from the issue of disasters and emergencies is not only politically short-sighted, it also ignores totally the causes and the effects of such phenomena.

Natural hazards such as hurricanes and earthquakes may be impossible to prevent. But they only become natural disasters if people are vulnerable. Why is it, for example, that an earthquake in Khilari, Maharashtra that registered 6.9 on the Richter scale, killed up to 35,000 people, when an earthquake of almost the exact same magnitude in Los Angeles in 1994 claimed only 57 lives? By reducing poverty we can help increase the coping capacity of vulnerable populations. Therefore helping people lower such vulnerability is as much a development issue as the environment, or women's participation in development. Moreover, the repercussions of natural disasters go for beyond the immediate casualty list that so transfixes the media. Secondary and longer-term effects can be equally if not more devastating. And they must be taken into account by developing practitioners.

It has been estimated, for example, that the damage to Mexico city's infrastructure from a massive 1985 earthquake amounted to US$3.6 billion. Yet over the subsequent five years, the negative ripple effect on that country's balance of payments resulted in a loss of $8.6 billion. Furthermore, reconstruction requirements forced Mexican authorities to revise their economic policies to meet an increased demand for public funding, credits and imports. The priorities for public expenditure were redirected to reconstruction projects, leaving many of the pre-disaster problems of the city and its people unattended.

In Bangladesh, floods in the recent past 2,000 people. But on closer examination we find that the toll was much more expensive than that: in each of these years the country's economic growth rate was halved by the delayed planting of rice and the destruction of seedbeds in the floods, further undermining the country's food security. All of these are consideration that go beyond relief, but they must be taken into account by development professionals.

Other emergencies may be more complex, but must be subjected to the same analysis. As the situations in Angola, Burundi, Somalia and the former Yugoslavia demonstrate, we know little about the dynamics of emergencies that arise from civil conflict. We do know, however that their cause usually lies in a lethal mix of poverty, poor governance and ethnic or religious rivalries exacerbated by profound social inequities. We are also learning that their resolution frequently requires the application of peacekeeping and political measures, combined with relief and development. Among the most virulent effects of such complex emergencies is the massive displacement of people; women and children are the principal victims, constituting 70 per cent of the world's refugees.

These complex emergencies around the world could easily get worse before they get better. This being said, carefully designed development efforts carried out as building blocks to national reconciliation in the fragile post-conflict stage-will need to increase commensurately. The appropriateness and the sustainability of these development efforts will be one of the most important factors in determining whether peace itself becomes sustainable. For example, the absence of carefully tailored reintegration strategies for demobilized soldiers and their host communities would be an almost open invitation to resumed violence.

Yet we must also be conscious of the impact of aid and try harder to prevent the need for relief in the first place. An increasing body of evidence suggests, for example, that emergency aid can sometimes be counter-productive in the longer term, increasing the vulnerability of populations and impeding recovery. Ironically, we find ourselves in situation today where it is far easier to obtain funds for maintaining refugees in their places of asylum than for helping them reintegrate into their own societies. In such cases, we may very well be helping to perpetuate the problem that we sought to relieve, as the presence of large numbers of refugees is sometimes itself a cause of conflict.

So how are we to proceed? And what exactly is the nature of the relief to development continuum that remains logical in the abstract but elusive in reality? The concept of a continuum does not imply a linear and absolutely progressive set of responses. On the contrary, it means that we are dealing with a set of processes rather than rigidly defined steps. It also means that development must be very much part of the disaster management process, and that the aim of the continuum must be to move from relief to rehabilitation and resumed development at the earliest opportunity. However, this resumed development must include conscious measures to reduce the vulnerability that caused the disaster or the emergency in the first place.

In other words, we must give greater thought to prevention before we reach for the "cure"—for humanitarian, political and financial reasons. (The Japanese insurance industry spends $200 million a year on disaster education alone). And as development practitioners, we must reconcile ourselves to the vastly more complicated environment in which we have to operate.

This means, for example that we will have to begin examining whether the economic policy "medicine" often prescribed will reduce conflict or enhance it. We will have to ask ourselves if the reconstruction period following a civil conflict or natural disaster is the right time to advocate cuts in social spending, as has happened in certain countries in Africa and Latin America. Similarly, is it really in children's best interests to build a school in a seismic zone without first ensuring its structural stability? And does it really make sense to urge drought-prone countries to increase their reliance on cash crops, as has been done in some instances.

A story that never made headlines anywhere involves hundreds of the poorest people in Bangladesh, whose homes remained intact during the floods of 1988, when many others were simply washed away. These people were fortunate enough to have obtained credit through the Grameen Bank for construction materials as well as instruction in the building of flood-resistant homes. The Grameen revolving fund had received start-up capital from International Financial Agencies. Since that time the effort has been expanded, and more than 10,500 flood-resistant homes have been built in the last two years.

This is just one example of the kind of action we need more of in fairly predictable and recurring circumstances such as the floods in Bangladesh, as well as in the more complex, man-made emergencies to which we must respond.

39

CRISIS AND NEW ORIENTATION OF DEVELOPMENT POLICY

The poverty in the South, the dislocation in the East, and the orientation crisis in the North are not isolated phenomena. Rather, they represent an alarming amalgamation of dangers that are globally interlinked.

The low effectiveness of international economic and development policy is rooted in two outdated paradigm on which the present worldwide strategy of economic development is based, namely that:

1. The western social and economic model optimizes the activation of productive forces—independent of the development state of a country and its culture and therefore is best suited to satisfy basic needs.
2. It is possible to launch the developmént of a society from the outside within a few decades—without regard to its cultural and historical background—through external input of money, goods, technology, expertise, and personnel.

The twin paradigm of the timelessness and transferability combined with cultural ecological, and financial restrictions—have led international cooperation and development down the wrong path.

Only if we acknowledge the true dimensions of the global dangers, if we recognize the limitations and shortcomings of existing political instruments, and identify outdated theories and contradictory special interests, can we outline the cornerstones of a new policy of global cooperation.

Cornerstones of a New Development Policy

Starting with critical review of the shortcomings and paradigms of the prevailing development strategy, the following ten cornerstones of a new development policy are offered for discussion:

1. *Broaden the concept of development*

Whether a society is considered developed depends on the size of its per-capita Gross National Product (GNP). Accordingly, the world is divided into a developed, semi-developed, and underdeveloped world. The yardstick for development, which has become the norm in the industrial countries, is one-dimensional. It only measures the monetary value of goods and services that are exchanged in the marketplace. This standard is too narrow economically because it compresses the multitude and complexity of cultural, societal, historical, social, and human values into a single economic category.

At the most, there can and should be agreement on what development and progress should not bring about: Inability to find enough work to meet the most basic needs; exploitation and oppression of people; loss of cultural wealth and institutions; destruction of natural resources. These, however, are the very values that are sacrificed by the prevailing development strategy. In the future, development policy must do all it can to stop the loss of skills and self-reliance, the plunder of natural resources, the erosion of cultural values, the violation of human dignity and human rights. Initiatives must prevail which are orientated on these values, and not just on the GNP.

2. *Concentrate development strategy on the internal potential of developing countries*

There must be an end to the manic fixation of development strategy on external inputs and external markets. A new development policy must, above all, improve internal conditions for a productive economy, promote domestic production factors on a broad basis, protect cultural and natural resources, and greatly increase the domestic supply of basic goods. Wherever external inputs are unavoidable, credits must be strictly tied to the productivity and the ability of a country to absorb transfers. External transfers should be concentrated on "software" for health, education, social participation, administrative, and legal jurisdiction. Such an approach could also promote training and indigenous technologies, which are so important for economic development.

The set-up and expansion of the productive sectors must be decided, planned, and implemented by the developing countries themselves, and they must assume full responsibility. The external pressures, which force the developing countries into full integration with the world market, must be removed. This presupposes a structural reduction of interest rates.

3. *Make development policy a central feature of politics*

Development policy must take the lead in mobilizing the various political forces and government departments to join the fight against the growing global dangers. It must ensure that the actions of all political departments are compatible with development policy is possible only if it becomes the central task of all political sectors, comparable to social and environmental policies, and the central goal of all policies. If development policy is to become a central task, development problems must become a priority in parliament and government. Society must understand that it is in the national interest to accept great global responsibilities.

4. *Reform the world economy*

The industrial countries must abolish their protectionism in agriculture as well the processed goods sector. Simultaneously, the developing countries need to be protected selectively and of a limited time against imports from the industrial countries. The undifferentiated structural adjustment policies imposed by the IMF must be revised. The trend toward regionalisation of the world economy should not be opposed; rather, in the interest of both South and East, it must be regulated constructively to form a new, regionally based world trade structure.

A reform of the international finance system is urgently needed. Interest and exchange rates should not mirror the national interests of the big industrial states and the special interests of large banks and venture capital. Rather, they must reflect the global interest in monetary stability lower and stable interest rates, and sufficient development financing.

However, strengthening the international financial institutions is in the global interest only if the countries of the southern and eastern hemispheres are allowed to exert some influence. An international financial court must guarantee that violations of strict regulations to ensure international stability and solvency can be protested in a court of law.

5. *Redesign the industrial society*

As a global social and environmental policy, the new development policy must induce the industrial countries to give up their excessive consumption of air, water, soil, resources, and space. Increased utilisation of energy-conservation measures and environmentally friendly technologies is overdue. The economic and social policies of the industrial nations must promote balance rather than growth. This requires radical changes in traditional economic thinking, habits, structures and processes.

In view of limited world resources, unsatisfied existential needs in South and East, and continuous population growth in the South, the only premise for the future can be—Growth rates in the South must be higher than in the North, but they should no longer be induced primarily by growth in the North. If economic policies continue to call for the North to provide the locomotive, the North will have to continue to acquire more resources than the South.

The North must relinquish the remaining growth frontiers to the South and East. The South must use this opportunity to activate its internal dynamic potential rather than integrate its economy with the North. However, ecological and social controls must be established at a much earlier stage than was the case in Europe.

6. *Strengthen development cooperation*

The share of official development assistance as a percentage of GNP, which dropped from 0.48 per cent in 1982 to 0.34 per cent in 1995 must be gradually raised again and reach at least 0.7 per cent in the year 2000, a goal which OECD established as early as two decades ago and which was reconfirmed at the Rio Earth Summit.

However, we must not succumb to the illusion that a doubling of ODA funds will even remotely meet the financial needs of South and East. State development policy must use its scarce public funds more effectively in the future. It must use restraint whenever partners in the developing countries can accomplish a task on their own and private initiatives and private enterprise are more competent to do the job. The government should be directly engaged only when it can be relatively more productive. Otherwise, it should limit itself to subsidizing private organisations.

7. *New orientation for development cooperation*

The state and its implementation agencies must abandon all direct responsibility for any projects, which require unbureaucratic action, economic efficiency and long-term productivity. It must make a much greater effort to involve NGOs and private venture capital in development projects. At the same time, the state must insist and guarantee that private actions are compatible with social and ecological concerns.

In the future, the main thrust of government projects should be the promotion of the internal potential of a country. This comprises the political and administrative framework conditions of a humane, socially and ecologically sound development. Constitutional government, social institutions which facilitate broad participation of the population in politics, society, and economy; efficient savings, credit, fiscal and financial systems; mechanism for income, property, and land distribution which promote productivity, justice, and social peace. In addition of this "Software" of development, the following is needed: A regimen for the protection of resources and environment; measures to prevent the short-term sellout of natural resources, elementary and general education and training, health care and social safety nets; capacities to develop science and technology.

8. *Reduce the debt service and activate private capital*

Public funds must be used to a greater degree for the financial rehabilitation of highly indebted countries in South and East; external demands for interest and principal payments must be adapted to the economic capacity of the respective country and its ability to execute external capital transfers.

Within the framework of international insolvency regulations, initiatives must be developed as a condition for the continuance of the present rules for write-offs which ensure effective cooperation from the banks and alleviate the banks and alleviate the heavy burden of private credits, with their high interest rates.

State development policy and private business interests should supplement each other. Government promotion of private enterprise initiatives for exports, investment, and employment in the developing countries must take into account their compatibility with development. In reverse, private engagements, which effectively promote development, must be actively supported by the government. A separate line item must be established in the development budget for such activation of private capital.

9. *Set regional priorities*

State development cooperation has been scattering its scarce funds not only among too many sectors, but also among too many partners. In the future, public funds must be concentrated regionally. More emphasis must be placed on regional programs, and development cooperation with threshold countries must be enhanced. A portion of public funds should be set aside to provide an incentive for threshold countries to assist the poorer nations in their own region as well as deal with poverty in their own country.

The new development policy could then also help lessen ethnic-national conflicts and promote peace by sponsoring regional cooperation in joint development projects. For this purpose, regional development funds must be set up for cooperation in the transportation, energy, trade, and finance sectors and last, but not least for regional security systems and disarmament. Such regional funds could also provide the means to project refugees and improve their prospects for an eventual return to their homelands.

40

AID EFFECTIVENESS AS A MULTI-LEVEL PROCESS

Parallel to the widespread decrease of aid resources provided by donor countries to developing countries in recent years, debate and research on how to make aid more effective has become a major concern. Usually, it is suggested that decades of development assistance have at best produced marginal results in terms of improving development levels in the South. Little mention is made of donor's policy shortcomings and the negative impact of these on efforts aimed at reforming and redefining development cooperation in order to enhance aid effectiveness. The policy parameters and operating frameworks of existing aid policies continue to inhibit higher degree of aid effectiveness. In many donor countries, opinion polls indicate waning public support for development aid.

Increasingly, the moral case for aid is called into question and deeper world market integration tends to be seen as the panacea to continued economic decline and social destabilisation in the South. Against this background, cooperation between donor and recipient actors is faced with a duel uphill struggle. First, fewer resources can be mobilized to meet growing developmental needs. On the other hand, to organise and manage development policies and programmes in a result-oriented manner, grows more difficult. The threat of further aid cuts and of further drops of public support for providing aid become ever more real. A closer look at the organisational complexities and political constraints under which development cooperation is expected to perform effectively may help to improve current aid management approaches.

Towards Conceptual Clarity

At first sight, catchy definitions of what constitutes effective aid might appear attractive to use, in particular with regard to economic indicators. The term "aid effectiveness" is easily used in the same vein as "efficiency", "significance" or "impact" of aid. At times, obsession to measure and demonstrate the results of aid supported development processes can be observed among policy-makers and administrators on the donor side. Still the understanding of aid and its effectiveness as being part and parcel of a cooperation relationship between donor and recipient side parties, is scarcely embedded in practice. To determine how to make aid more effective requires more than a quick impact analysis of an individual and perhaps even isolated development project. Consequently, defining the concept of aid effectiveness needs to take into account at what levels cooperation is focused on. To strive for sustainable and effective modes of development cooperation will entail the need to combine recipient ownership of the development process with donor accountability concerns.

Performance expectations cannot be exclusively placed on the recipient while donor interests, their aid management systems and procedures remain unchanged.

An extended and more analytical, process-oriented definition should take into account four main aspects of aid effectiveness:

(a) Effective aid must relate to the building and/or strengthening of in-country aid management capacity,

(b) To maximize the degree of aid effectiveness, local ownership of the aid process is essential from setting of priorities through policy formulation and implementation on to the evaluation stages of the process;

(c) Increasing recipient side capabilities to take charge of aid relationship, will need to be combined with arrangements to meet legitimate donor accountability concerns;

(d) Aid effectiveness is a two-faceted objective: its realisation is equally dependent on increased transparency of donor motives and on dropping of nondevelopmental, political and economic aid objectiveness of donors.

In addition a broader range of stakeholders in the aid relationship needs to be actively involved: extending beyond accountability government and implementing agencies, to include democratic institutions and organisations of civil society and of the private sector.

Applying any definition of aid effectiveness without disaggregating macro-economic data and taking into account country specificity will only lead to unhelpful generalisations about aid and its effectiveness. It would seem more appropriate to adopt working definitions against which to assess effectiveness of aid resources at a country-specific level. On such a basis one could expect to arrive at more reliable indicators of how well aid resources contribute to improving developmental standards and meeting existing needs.

From Definition to Success—Key Requirements

Having reached agreement between the recipient and donor on what should constitute effectiveness of aid is only a starting point. Embarking on democratic, peaceful and participatory patterns of economic and social development must follow: to arrive at significant and lasting improvement in many of the least developed countries will be a long-term process. This being said, it is crucial to design and implements such forms of development cooperation which involve a wide range of recipient side actors, not only from the government side but also from civil society at large. Seen as a process of increasing inclusion of intended beneficiaries of aid, the commitment to decentalize as well as entrust aid and its management grows in importance.

To fully capture Third World development realities, policy frameworks inspired by neoliberalist-type of development concepts and theories are grossly inadequate. The views and positions on aid articulated in the World Bank and the IMF, or in many if not most bilateral aid administrations in OECD countries, represent only one side of today's international cooperation, namely the donor side. The major weakness to point out with respect to this locus of debate, is a profound under representation if not even a total absence of recipient experiences and perceptions on aid in general and on its effectiveness in particular. There should be little doubt that ignoring to not actively identifying and involving such perceptions, leads to strongly donor driven aid.

To circumvent recipient side insights and views on strengths and weaknesses of aid strategies and mechanisms, will result in limited local commitment and sense of ownership over the aid process. Mutual decision-making between donors and recipients remains a rare policy approach. Aid procedures that are based on local management and less control-oriented donor roles in the aid process are still exceptions in development cooperation.

Structurally, in terms of the policy environment within which development aid is expected to function, the overriding policy framework is generaly based on structural adjustment policies (SAP). But the underlying conclusion made by proponents of SAPs that these policies induce aid effectiveness, has yet to be proven valid. It must suffice at this point to emphasize that there is no a priori relationship between world market integration under structural adjustment and sustainable development in poor countries. Aid to these countries which is solely intended to reinforce fundamentally uneven and unequal patterns of world market integration should be scrutinized critically.

Some central issues need to be addressed in the course of improving aid and its effectiveness:

- institutional dimensions of aid relationships require strong policy-attention, both on the donor and the recipient side;

- capacities to effectively identify and formulate aid priorities need to be strengthened in recipient countries;
- local capacities to sustain reform efforts must be reinforced.

Levels of Intervention

If the design of aid and the terms upon which it is provided to a developing country are largely determined by the donor, the aid relationship can be characterised as essentially hierarchical. Recipient side views will rarely surface, as they are either not identified, or not well formulated. Possibilities of a recipient-led development strategies can be limited. Unless scope is provided to the recipient side actors to assume responsibilities, aid effectiveness is likely to remain low or fluctuating, and the sustainability of donor aid efforts will remain doubtful.

National planning processes and courses of national development in recipient countries should be seen as most effective where they are led under local responsibility and control. To arrive at this ideal situation, gaps need to be reduced and closed at the various intervention levels.

Donor aid resources provide valuable support for this process. Their effectiveness in meeting long-term objective of aid will need to be assessed on the basis of how well they perform at the different levels. Individual donors will expectedly perform differently at the various levels. What will prove to be the ultimate test for effectiveness is how well the donor aid performance accomplishes the broader objectives of development cooperation and how well it includes sustainable results.

In the analytical frameworks outlined here, development cooperation would seem to be confronted with the effectiveness gaps at the:

- structural level: International trade and investment patterns, debt problems and world market integration process appear as long-term constraining factors upon aid and its effectiveness;

- policy level, dialogue and partnership in development cooperation are instrumental factors in recluding planning and co-ordination gaps with regard to policy analysis and formulation;
- institutional level is where pertinent capacity gaps exist: capacity development efforts of donors and technical assistance measures play an important role in addressing weaknesses in aid effectiveness within a country's institutional setting;
- finally, at the level of aid projects (programmes), it is generally the lack of sustainability of aid interventions which causes development activities to falter once donor support decreases or stops. In addition to technical cooperation, financial and material inputs serve to maintain project momentum and goal realisation. The issue of how to develop local capacity sufficiently in order for indigenous organisations to continue project activities initially supported by donor aid, remains the most important issue to address at this level.

Fostering Aid Effectiveness

Donor and recipient development efforts are too often isolated from one another, or poorly coordinated. They fail to address managerial and implementation bottlenecks. Cross-sectorial linkages, as well as interdisciplinary approaches to aid problems are only slowly gaining ground. It is increasingly obvious, that decisions on aid issues are subjected to concerns outside of the responsible ministry: finance ministers, and unfortunately even defense ministers have a strong say in how much aid is to be provided, where it is to be concentrated and under what terms to be utilised. Inside of recipient countries, large portions of national budgets are allocated to non-development priorities with little or no impact on alleviating urgent poverty problems.

Development cooperation may make the biggest impact and be executed most effectively where donors and recipients agree upon multi-level aid strategies. To give an example: building a road to a remote rural area may well be done in an effective project manner: it is equally important to have a functioning transport authority in place to ensure maintenance of the roads. If this authority operates within a nationally defined infrastructure policy, best in accord with national trade and investment priorities, then the effectiveness of the project-level road building programme has a good chance of being high.

Institutional changes to set the stage for a profound reform process in development cooperation are needed. Reprioritising national budgets to reflect identified in country development needs may be one step. Setting up policy evaluation and formulation units can be complimentary measures. Deregulating markets and investment rules may serve to please donors, but dumping of cheap products which strangle local production efforts may easily result. Regional cooperation, including intensified South-South cooperation can provide some counterbalance. There are only a few areas where changes in the current system of development cooperation can occur, with a view to better manage the complexities of aid and the social, cultural, economic and political backgrounds against which they take place. The will and commitment to take policy action in both donor and recipient countries, through the broadest range of stakeholders and institutions as possible, will be the test for genuine efforts at improving development relations between North and South and organising cooperation effectively.

THE DEMATERIALISATION OF THE WORLD ECONOMY

The first Industrial Revolution marked the transition from robber—and—plunder colonialism to the systematic development of the "overseas" territories in the framework of the international division of labour between raw materials suppliers and manufacturers of finished goods. There was an "historic integration" of the colonised areas in the development of their parent—states. What will the third Industrial Revolution do for the Third World? Will it now come to an "historic separation"?

The end of the East-West conflict was reason enough to talk about a radical change in world politics. But at the same time an upheaval in the world economy is taking place that possibly will have even wider impacts. As a reference point for the following thoughts, three dimensions of this change are pointed out:

1. The upgrading of processing information rather than materials as object of economic activity (technological dimension);
2. The evolvement of global communications networks (sociocultural dimension);
3. The change of the nature of work (socio-economic dimension).

All three dimensions can be summarised under the buzzphrase "tertialisation of the world economy."

In that respect, talk of the "Third Industrial Revolution" is misleading. It is not about a third epoch of industrialisation, but about the beginning of a de–industrialisation, the transition from the industrial to the information society.

Historic Separation?

In the 1960s and early 1970s, there was often talk of the Third World as the Third Sector of the world economy. Also then the Third World was not much more than an "imaginary community". But as such it had a certain significance in world politics. This implied not only its strategic role in the East-West conflict and its ideological function as the supporter of different "third paths" between capitalism and socialism. It was also about the Third World's attested "chaos power". That linked the fear (in the North) and the hope (in the South) that the developing countries would be in a position to cut off the industrial nations from supplies of important raw materials, thus putting them under pressure. But it was soon seen that both sides had over estimated this possibility, even with regard to oil. Instead of supply bottlenecks arising, raw materials prices plummeted. For some commodities, the fall in prices exceeded those of the Great Depression of 1929/30.

This was due, *inter alia,* to the conjunction of lower demand from the industrial nations and expansion of production by the raw materials suppliers. Business activities dependent upon the supply of raw materials are tending to lose importance compared with the overall development of the global economy. The reason for this is to be seen in the transition from a material to an information economy.

This transition is taking place in line with the revolutionizing of data transmission and the expansion of financial transactions which are not directly related to changes in the production of materials. The speed of the changes is remarkable.

However, the dematerialisation of business activities does not lead to decoupling of the Third World from the world economy. Declining market shares in world trade are not the expression of separation, but a loss of the affected countries positions in the world economy. Thus, the impact of dematerialisation is "only" that the negotiating positions of raw materials suppliers *vis–a–vis* the industrial nations will deteriorate further.

Differentiation of the Third World

But the radical change in the global economy is affecting some developing countries worse than others. Sub-saharan Africa, and some countries in West and South Asia and Latin America are being pushed back further. The oil-producing countries with their high per capita export earnings will be able to hold their positions in the world economy for some time to come. The threshold countries of East and South-East Asia can expand theirs so long as they can continue to attract a growing share of global industrial production, and at the same time participate in the tertialisation of the world economy in the shape of rapidly growing financial transactions. Thereby it should be noted that the degree of tertialisation in itself is not an adequate indicator for economic avant-gardism. Brazil exhibits a high degree of tertialisation in combination with a low macro-economic development dynamics. A good part of its tertialisation is being achieved by speculative financial transactions with their inherently greater risks and uncertainties than in the industrial countries. Such dangers have been demonstrated by Mexico's peso crisis and its repercussions on the whole of Latin America.

In some Third World countries, a "location annuity" has replaced the old raw materials one. Here it's about providing locations for off-shore transactions which offer international capital traders a maximum of freedom of movement combined with low taxation. Suitable for such operations are small countries which, despite low levy rates, achieve significant income in macro-economic terms.

The radical changes in the world economy are spurring the differentiation of the Third World without, however, necessarily fostering a dissolution of the Third World as an "imaginary community". It is precisely the advanced countries of East and South-East Asia that are showing a certain interest in the formulation of joint positions of the "South" in order to secure their own positional gains in the global economy. It's not by chance that the non-aligned countries and the Group of 77 have formed a joint coordination committee, and that the ASEAN countries are changing course on the international human rights policy.

Hitherto, the developing countries' strategy was to broaden the concept of human rights as a justification for demands on the industrial nations. But of late some developing countries, led by the ASEAN states, have questioned the universal validity of human rights even after their universality was confirmed by consensus at the Conference on Human Rights in Vienna in 1993. Playing a role in this policy is the governments' fear that due to the expansion of global communications networks, the behaviour patterns and preferences of their own people could in some way become similar to those of the West. As the rulers see it, that would be detrimental to the continuation of the development models practised so far.

Internet Creates New Cultural Dimension

Much information which Asian governments view as subversive in already globally available on the Internet. The old struggle over the world information order, which at first was primarily a clinch between East and West, is thus taking on a new dimension. For with the growing importance of computer literacy to a country's ability to assert itself on world markets, the Asian threshold countries have not only an interest in controlling the on-line communication but also to expand it and the know-how that it requires.

Even the critics of any interventions in the internet and other global communications networks must admit that modern communications technologies are politically blind and their use in itself does not represent progress. The setting up and expansion of global information highways will offer forum not only to people who want to use it for education and enlightenment, but also to all shades of fundamentalists. These highways will not necessarily bring the misery of many Third World regions closer to the industrial countries, but possibly rather strengthen the tendency to process all world events as entertainment.

Global Two-Thirds Society

The gravest aspect of the current upheaval in the world economy is its negative impact on jobs. The information economy needs for fewer workers than an economy based on materials. Instead, the demands on the skills of the workers are growing. Twenty per cent of the world workforce will in future be employed as (overworked) "intelligence workers". Eighty per cent will work part-time, if they are not underemployed or jobless. So the tertialisation of the global economy delivers more underemployment rather than more leisure time. The workers who are rationalised out of their jobs in the industrial sector cannot be absorbed by the service sector because it, too, is not left untouched by rationalisation measures. The civil service is also cutting back on staff. At all levels, there is a race to make the greatest possible savings on payrolls. At the same time, there is growing pressure to cut costs in providing for the victims of this development. That means thinning out the social security safety net.

The bottom line is that the two-thirds society, which developmental action groups hitherto assumed was limited to the Third World, is spreading worldwide. That, however, will not in the foreseeable future lead to an amendment of the North-South disparities. It is true that the change in the global economy is taking place faster and to a greater extent in the industrial nations. But rationalisation is also happening in the developing countries in a bid to boost their competitiveness. So the upheaval in the world economy aggravates the problems which exist in a majority of the developing countries, while creating new ones in the industrial nations. The need for action on the North-South policy is growing, while the industrial nations. The need for action on the North-South Policy is growing, while the industrial nations' scope for concessions and compromises is shrinking. The new social question which is now crystallizing at global level is not being answered. The consequences are unforeseeable.

Another Loser?

Its more probable that a sharpening of the North-South confrontation is to be reckoned with. For the industrial nations will attempt to keep the social costs of the information economy at bay for as long as possible. The trade unions will thereby compete with the developing countries for jobs for their members. But this policy has its limits precisely because of the peaking of the problems in the industrial nations. Overstepping these limits means war, and passively accepting them chaos and social decay. Solutions could be sought in two directions: effective taxation of the information economies, and the creation of jobs in the non-profit sector. But it is possible, there are no global solutions for global problems. That would mean for at least part of the Third World a renewal of the old debate on partial decoupling from the world economy.

42

WHAT'S DRIVING MIGRATION

The scale and diversity of today's migrations are beyond any previous experience. Rapid urban growth and environmental degradation in rural areas have led to internal migration affecting hundreds of million of people. Migration is now seen as a priority issue equal in political weight to other major global challenges such as the environment, population growth and economic imbalances between regions.

Families and households form the basis for economic growth, social development and personal fulfilment. Decisions, by individual women and men on marriage, family, a place to live, shape the destinies of communities and nations. National policies and international conditions provide the context for individual decision-making. Effective development policies, including population, reproductive health and family planning policies, address this reality.

Data on national and global population trends set the agenda for national policy. An important element of population programmes is gathering data that will allow policy-making responsive to the realities of daily life, and to the needs and aspirations of individuals.

The dominant feature of global demographics is still growth. Age distribution is a growing concern, as the numbers of young and elderly people, grow, relative to the working-age population. The world is growing steadily more urban. From being a sign of strength and dynamism in the national economy, the rate and scale of urban growth has become increasingly a cause for concern. The influx of migrants to the biggest cities may be weakening both urban and rural sectors.

International migration is small in extent compared with internal movements, but has a disproportionate impact. Both internal and international migration are driven by population growth, and by inequities between countries. Migration is one of the choices which shape people's lives and the destiny of nations. But it can also be a symptom of inequity and underdevelopment. Migrants are by definition the most vulnerable members of the host community. Their living and working conditions should be protected.

Open and frank exchange of information and views between host and sending countries is needed more than ever. The aim of the international community should be to protect the right to move, but to ensure that movement is voluntary and that it stimulates rather than holds back personal and national development. "The point of departure should be the human right to live and work where one pleases, so long as it does not infringe on other people's rights to do the same."

The Urban Transformation

The rural sector is declining in importance and its contribution to national economies. It is increasingly part of a unified economy based on the city. Contact with the urban areas is easier than ever and is encouraged by rural development.

Temporary and circular migration is giving way to more permanent settlement. The largest cities are under increasing strain, and residents are encountering increasing difficulties in improving or even maintaining living conditions. Nevertheless, migration continues, driven by a variety of forces both positive and negative. The choice to move can be part of a strategy for survival or personal development; but it is often enforced by external conditions.

The urban transformation is irreversible, but the rural sectors must also be strengthened to balance the developing economy. Attention to gender issues will be crucial in ensuring a successful transition. The forces driving internal and international migration have much in common. Demographic pressures are contributing to both. As the pressures encouraging migration increase, the options for migrants become more limited. This collision is contributing to the atmosphere of crisis surrounding both urban and international migration.

Costs and Benefits

Migration is the result of individual or family decisions. But it is also part of social process. In economic terms, migration is as much a global phenomenon as trade in commodities or manufactured goods. It is part of a broader pattern, and evidence of changing economic, social and cultural relationships.

But migration may be evidence of a different kind of relationship: the combination of poverty, rapid population growth and environmental damage is a powerful destabilizing factor driving urban growth and eventually international migration. On the recipient side, migration has usually been seen as evidence of a thriving economy today's industrial states were built in part by migrant labour, skills and investment. In today's increasingly uncertain conditions, migration may be seen as a threat to the security and well-being of the local workforce and society at large.

The only effective means to reduce migration pressures over the long term are to slow population growth; to stimulate economic growth and job creation at home, and promote the development of the individual and the family as the basic economic and social unit.

A Question of Gender

It is often assumed that most migrants are men, in reality, women make up nearly half of the international migrant population. Gender differences in social and economic roles affect migration decision-making, household strategy, and the sex composition of labour migration. Attention to the gender dimension of migratory movements ought to be an important component in population and development planning.

Women frequently take the initiative in migration decisions, which may reflect limited opportunities in rural areas. Low status limits women's choices at home and may increase pressure to migrate, but it may also affect life in the host community. Opportunities may be limited by lack of education or skills, or by customer limitation on women's freedom of action outside the family or ethnic group. Paid employment for migrant women is usually in the lowest wage, lest secure, and lowest status jobs, mostly in housework, child care and trade.

Most educated women end up in the same low-status, low-wage production and service jobs as unskilled female migrants. Men too, experience downward mobility, but the contrast in the decline in women's employment status is far greater. Despite these disadvantages women migrants have become significant economic actors. Their status may be improved by migration, but the advantages are not clear-cut. Women's status as migrants is affected by their vulnerability, and by their lack of reproductive freedom. To ensure improved status they will need both legal protection and essential services, including reproductive health services.

Refugees

Refugees in the 1990s are overwhelmingly in Asia, Africa and Latin America. Their numbers are large, about 17 million, and growing rapidly. A further 3.5 to 4 million were thought to be in "refugee-like situation", though estimates are probably extremely conservative, and an estimated 23 million people internally displaced.

It is important to recognize the common roots of refugee and other forms of mass movement of populations. At the same time, despite the difficulty of distinguishing between political and socio-economic causes of migration, there is a clear need to distinguish between refugees and other groups of migrants. Participation in international efforts of burden-sharing would ensure that most refugee problems would be dealt with in their regions of origin.

Conclusions and Policies

Migration highlights linkages and interdependencies within countries, with many implications for development agendas, including population programmes and development assistance.

Policies to regulate or moderate international migration have concentrated largely on urban growth. They have been only intermittently effective. The most successful have concentrated on stimulating rural development and the growth of alternative urban centres

Migration is also a personal of family decision, which is affected by external conditions such as poverty or environmental degradation, improving conditions of personal and family life can make a crucial difference in the decision to migrate, reducing dependence on migration as a strategy. Because migration is the result of personal and family decisions, it can be influenced by policies that improve the quality of life.

This offers the opportunity for policies emphasizing individual development, among them education, health (including reproductive health) and family planning. Such policies are particularly relevant to the strategies must take into account gender differences in social and economic life and the differential effects of policies.

Migration decisions are about family security and long-term-life-chances, rather than simply the maximisation of income. They are ultimately strategies designed to look after the individual's and the household's needs, safeguard their security, and respond to their aspirations. If the goal is to reduce migration pressures through development it will be essential to increase the capacity but reduce the need to migrate. Long-term external support will be required to make such policies a reality, particularly in areas of rapid population growth and potential mass outward flows. Highly co-ordinated allocation of development assistance can help establish priorities and focus attention on basic needs. The challenge to both international donors and co-operating governments is to direct programme spending to the areas where it can be most effective.

43

NOT YET FOSSIL FUEL

People were gathering wood for their fires more quickly than it could grow back. In India, wood was being burned 50 per cent faster than replacement tress were growing. Among environmentalists, the perception became widespread that village cooking fires were consuming the Indian forest.

Firewood demand was a minor cause of deforestation. People were mostly using twigs and dead branches for fuel, leaving trees standing. The main perpetrators of deforestation were not women preparing for the evening meal, but farmers clearing land for crops and livestock. To this day, however, the belief persists that fuelwood scarcity drives deforestation. That popular misconception has hampered efforts to address serious fuelwood problems that do exist. While not quite deforesting the globe, these problems are undermining the well-being of millions of people in India.

Until recently, most biomass consumer lived in rural areas. As populations have grown, and the number of trees has diminished, searching for fuelwood has indeed become a demanding task. In some areas of India, for example, collecting firewood was a two-hour task only a generation ago, yet today it is almost an entire day's expedition—every day. That constitutes an enormous erosion of productivity in other kinds of work. And the problem is worsening rapidly.

What makes the situation even more difficult is that fuel-wood problems have now spread from rural areas to the cities. In the 1950s, the cities in India were relatively small, inhabited mainly by those who could afford such "modern" fuels as kerosene, liquefied petroleum gas (LPG), and electricity. But in the past three decades, urban populations have exploded as migrants from the rural areas pour into the cities in search of jobs and higher standards of living.

Despite the availability of the "modern" energy sources to some of the city dwellers, the majority of migrants cannot afford them. Wood remains their fuel. But instead of collecting it, they now must buy it from vendors.

Biomass traders search for wood in communal woodlots, and consequently procure lit as a "free" commodity. The price of he wood, which represents only the transport costs and trader's markups, excludes the production and replacement costs at the expense of rural people and the natural resource base. After decades of woodland exploitation in some areas of India, the shortage of firewood in rural areas has become so severe that villagers are compelled to use cow dung, dried leaves, and grass for fuel. The fragile soils of the farmlands are thus denied essential nutrients and organic matter.

Villagers are working hard to reduce the exploitation of their resource base by the urban entrepreneurs. For example, in a village not far from the capital city of Delhi, residents decided to charge traders for wood taken from their communal land. Extension agents were brought in to train the villagers in negotiating wood prices, maintaining accounts, and establishing an agro-forestry project. The revenue from wood sales is now used by the villagers to plant trees that will not only provide fuel for their personal use and for sale, but can be used for other purposes such as construction.

Dependence on wood for fuel is not just a challenge to economic sustainability, but also a threat to Indians health. In the homes of low-income families, where traditional wood stoves are widely used for cooking, adequate ventilation is often lacking. These stoves require large quantities of wood, are unable to retain heat for prolonged periods, and send much of the fuel up in smoke. Inhaled by women as they cook, the smoke has been identified as a major cause of respiratory problems such as bronchitis, and of damaged eyesight.

Improved wood and charcoal stoves are now for sale in India. Beçause they require significantly smaller quantities of firewood or charcoal, they are successfully curbing the wood demand, and hence promoting biomass conservation.

In the cities, as vendors travel longer distances into the hinterland in search of wood, and the transportation costs increase, more of the market is shifting to charcoal, which is easier to transport and more convenient to use in cramped urban quarters. However, widespread use of charcoal puts even grater stress on the environment because conversion from wood to charcoal requires twice as much timber to yield the same cooking energy.

In addition, once the conversion, transportation and combustion processes have been accounted for, charcoal emerges as one of the leading producers of carbon dioxide the most prevalent greenhouse gas. A meal cooked with charcoal produces three times more carbon dioxide than the same meal cooked with wood. On the other hand, charcoal can be made into a more efficient fuel by producing it in kilns that retain a higher proportion of the energy content within the charcoal. And the amount of charcoal used can be reduced by the use of more efficient charcoal stoves, which control airflow to the fuel and are insulated to minimize heat loss.

Looking for alternatives to fuelwood and its charcoal derivative, a number of communities are experimenting with solar box cookers and biogas digesters, both of which use renewable sources of energy and are pollutant-free and energy-efficient. The solar cookers and the digestors have been successful on a small and localized scale. The new methods have not been used on a big scale because of the high initial costs along with a variety of cultural biases and superstitions. For example, preparing and cooking food is an evening social activity, and solar cookers have to be used during the day. Biogas digestors are similarly constrained, since in some cultures the use of human and livestock waste, as fuel is unacceptable.

The fuelwood crisis is complicated one, and the easy, large-scale solutions that were originally recommended, such as the establishment of peri-urban plantations to increase wood fuel supplies, evidently will do little, if anything, to alleviate the problem. However, over the years, through mistakes and project failures, a few useful lessons have been learned. One such lesson—counter-intuitive though it may seem to many environmentalists—is that since biomas fuels will continue to play a major role for years to come, greater emphasis needs to be placed on finding ways to increase the woodfuel supply. Farmers, for example, can be encouraged to plant trees that will provide not only fuel but other products such as fruits, fodder, and lumber. And when the new high-efficiency stoves are made more widely available, the demand for trees will be reduced, even as their supply is increases.

BIBLIOGRAPHY

Anand, R.P., *Legal Regime of Sea Bed and the Developing Counties,* 1975.

Bhatt, S., *Environment Protection and International Law,* Radiant Publication, Kalkaji, New Delhi, 1985, p. 122

Bhatt, S., *Environmental Laws and Water Resources Management,* Radiant Publication, India, and Advent Books Inc. New York, 1986, p. 335.

Behrman, Danial, *In Partnership with Nature—UNESCO and the Environment* (Paris, 1973).

Bell, Daniel, "Technology, Nature and Society", *American Scholar,* Summer 1973.

Bentley, Glass, "Biology and Human Values", USIS, New Delhi.

Book of Nature: The Way Things Work, Published by George Allen and Unwin Ltd., 1981, p. 525.

Boulding, Kenneth E., "New Goals for Society", S.H. Schun, ed., *Energy, Economic Growth and the Environment.*

Carr, E.H., *What is History,* 1961.

Darlington, C.D., *The Evolution of Man and Society* (London, 1961).

Downing , Paul B., ed., *Air Pollution and Social Sciences* (New York, 1971).

"Drive to adopt national water policy", *Times of India,* 22 July, 1983.

Dubos, Rene, "Man and his Environment", *Britannica Perspectives,* vol. 1, 1968.

Einstein, A., *My Views,* ed. by S.K. Bandopadhyaya (Calcutta, 1976).

"Environment Research Programme", prepared by NCEPC, Department of Science and Technology, New Delhi.

Forbes, R.J., "The Conquest of Nature and its Consequences", *Britannica Perspectives,* vol. 1, 1968.

Fowler, John M., *Energy and Environment* (New York, 1975).

Fuller, Buckminister R., *Operating Manual for Spaceship Earth* (New York, 1969).

Gandhi, Indira "Poverty Greatest Pollution, says Mrs Gandhi", *Time of India,* 8 September, 1981.

Glenn, Seaborg, "Science, Technology and Development: A New World Outlook", USIS, New Delhi.

Hacoley, Amos H., *Human Ecology* (New York, 1950).

"India must Develop own Ecology", *Time of India,* 8 October. 1981.

Marion, Jerry B., *Energy in Perspective* (London, 1974)

Misra, K.C., *Manual of Plant Ecology,* New Delhi, 1980.

Mukherji, P.K., *Life of Tagore,* Trans, by S.K. Ghosh, 1975.

Mumford, Lewis, "The Future of Cities", in *Basic Issues in Environment,* E.J. Winn, ed., 1972.

Palmslierna, H., *Future Imperatives for Human Environment,* 1972.

Pavithran, A.K., "World Futurology", *Eastern Journal of International Law* (Madras), vol. 9.

"Plans to usher India into 21st Century", *Times of India,* 24 October, 1985

Polunin, Nicholas, "The Biosphere Today", *The Environmental Future,* Proceedings of Ist International Conference on Environmental Future in Finland, ed. by N. Polunin, 1972.

Radhakrishnan, S., *Recovery of Faith,* 1967.

Report on the State of Environment Prepared by Centre for Science and Environment, New Delhi, 1985.

Sarkar, Mahendra Nath, *The Cultural Heritage of India*, vol. 1.

Sen. Sudhir, "Blueprint for a better World", *Times of India,* 2 March, 1980.

The Limits to Growth, a report to Club of Rome (New York, 1972).

The Mind of J. Krisnamurti, ed. by L.S.R. Vas (Bombay, 1971).

Toynbee, Arnold, "Man and his Soul", *Hindustan Times,* 4 January, 1968.

United Nations List of National Park and Protected Areas, 1985.

Vivekananda, Swami, *Complete Works*, vol II (Calcutta, 1968).

Ward, Barbara and Dubos Rene, *Only One Earth: The Care and Maintenance of a Small Planet,* Report to UN Conference on Human Environment, Stockholm, 1972.

"Wildlife laws in India", *Times of India,* 4 March, 1985.

Ward, Barbara, *Progress for a Small Planet*, 1979.

INDEX

E

F

G

T

U

W

Y